Velvet Revolution at the Synchrotron

Inside Technology

edited by Wiebe E. Bijker, W. Bernard Carlson, and Trevor Pinch

For a list of the series, see pages 153–155.

Velvet Revolution at the Synchrotron

Biology, Physics, and Change in Science

Park Doing

The MIT Press
Cambridge, Massachusetts
London, England

for Mary Houser Doing, Pennsylvania State University, Class of 1928

QC
51
.N7
D65
2009

For information on quantity discounts, email special_sales@mitpress.mit.edu.

Set in ITC Stone Serif and ITC Stone Sans by Graphic Composition, Inc., Bogart, Georgia. Printed and bound in the United States of America.

Library of Congress Cataloging-in-Publication Data

Doing, Park, 1965–
Velvet Revolution at the synchrotron: biology, physics, and change in science / Park Doing
p. cm. — (Inside technology)
Includes bibliographical references and index.
ISBN 978-0-262-04255-0 (hardcover: alk. paper)
1. Physical laboratories—Sociological aspects. 2. Physical laboratories—New York (State)—Ithaca. 3. Laboratories—Sociological aspects. 4. Laboratories—New York (State)—Ithaca. 5. Cyclotrons—New York (State)—Ithaca.
6. Research institutes. 7. Science—Methodology. I. Title.
QC51.N7.D65   2009
306.4'5—dc22

2008041728

10 9 8 7 6 5 4 3 2 1

Contents

Acknowledgments

I thank my dissertation advisor at Cornell University, Professor Trevor Pinch, who allowed me to sit in on his class before I was officially a student, listened to me as I informed him that I wanted to write "one of those laboratory studies," and then proceeded to patiently and thoughtfully guide me through both my dissertation and the completion of this book over the course of more than a decade. For your scholarly insight, offered at just the right times and in the right way, your support and enthusiasm, and your friendship, I thank you. You helped me grow as a person and nurtured my work. For your interest in me, I am humbled and truly grateful.

I thank my parents, Park A. and Erica H. Doing. Your unflagging support and unconditional love are the engines that drive my work and life.

I thank my grandparents Anton and Eva Horwath for inspiration.

I thank my children, Ben and Georgia Doing, who spent many an hour whiling away time at the laboratory and asking me important questions about it.

I thank Kristina Doing-Harris, Mark Harris, Tony Doing and Sarah Mitchell Doing, Nick and Kristen Doing, Corrie, Ailish, Megan, Maddie, Katie, and Nicholas.

I thank Sarah E. M. Grossman for her belief in me as a person and a scholar.

I thank the Science and Technology Studies Department at Cornell University, including my dissertation committee members Ron Kline, Sheila Jasanoff, and Michael Lynch and my reader Peter Dear. Your intellectual support and challenges are an ongoing source of deep inspiration.

I thank Peter Taylor for a crucial happenstance afternoon meeting at the Big Red Barn that changed the course of my life.

I thank fellow graduate students Bill Lynch, Arthur Daemmrich, Alec Shuldiner, Jenny Reardon, Tracy Spaight, Suzanne Moon, William Wittlin, Sonja Schmid, Ayrin Martin, Samer Alatout, and Simon Cole.

I thank the philosophers of the laboratory: Jim LaIuppa, Alan Pauling, Chuck Henderson, Chris Payne, Brian Carrol, Chris Conolly, Walt Protas, Basil Blank, Qun Shen, Randy Headrick, Ernie Fontes, Ken Finklestein, Tammy Yee, Stefan Kycia, Bob Batterman, Don Bilderback, Bill Miller, John Kopsa, Jeff White, Tom Irving, Sol Gruner, Dana Richter, Ben Clark, Chris Heaton, Arthur Wall, Mark Keeffe, Peter Quigley, Lana Walsh, Keith Brister, and Karl Smolenski.

I thank Professor Boris Batterman for allowing this book, and my graduate studies, to occur.

I thank the staff of the MIT Press, including Sara Meirowitz, Erin Shoudy, Paul Bethge, Megan Schwenke, and especially Marguerite Avery, for their diligence and enthusiasm.

1 Birth of a Hybrid Laboratory

It is 1993. In the crystal cold hours before dawn, a campus sleeps as yet another winter night gradually unfolds into morning. Below the frozen soccer fields, however, under a thin layer of grass, 50 feet of dirt, and 2 feet of concrete, the buzz of activity is lively and constant. Electric fields oscillate. Magnetic fields hold strong and steady. Electrons orbit at nearly the speed of light inside a roughly circular evacuated metal pipe while their counterparts, positrons, pass them by, traveling just as fast in the opposite direction. X-rays emitted by the accelerated subatomic particles emanate at a tangent to the circle, traveling straight down pipes connected to the ring. Lights flash. Needles waver. Numbers blink on screens. The machine is on. That's what the laboratory members call it: "the machine."

In this book—a study of a present-day hybrid physics and biology synchrotron laboratory—I examine the relationship between technical knowledge claims and organizational modes of authority and control. Through an analysis of three episodes—one involving contestations over instrumentation development, one involving laboratory operations, and one involving methods of experimentation—I explore how laboratory members engage in "epistemic politics" whereby technical knowledge claims are

Schematic of the particle accelerator and tangential x-ray beamlines underneath Cornell's soccer fields. The laboratory building is at the bottom of the picture. (Cornell 2000)

implicated in modes of authority, access, and control. Further, I assert that the epistemic-political order at the lab changed in conjunction with the rise of protein crystallography in synchrotron science at the end of the twentieth century. Finally, I question what this change means for the epistemic status of scientific facts used in, and emanating from, the laboratory.

Around the machine, all underground, are control rooms, cavernous halls, offices, hallways, and lounges. In this labyrinth, scientists, technicians, machinists, operators, and administrative personnel work to produce and account for the output of the machine. Among these various types of staff members working in, around, and on the machine, there is an important distinction: one group—the particle group—is interested in the smaller particles produced by electron-positron "interactions" as monitored by their detector. The other group—the x-ray lab—is interested in how x-rays that emanate from the machine interact with various materials.

I want to bring out "what the lab is" as it unfolded for me and as it changed. I want to show how different conceptions of technical practice and technical practitioners at the lab were put to use at different times by various lab members in the course of their work. Considering how understandings of who can produce knowledge and how they can do so changed in the course of the technical life of the lab is a means of exploring the relationship between the contingency of scientific practice and the output of that practice.

As dawn breaks, the machine's operator arranges for the positrons and electrons to collide, all the while monitoring vacuums, temperatures, beam intensities, and beam positions. With regard to the particle collisions, records of the results from this night will be added to those of previous and future nights, months, and years in the hope of producing indications of the actions of a particular particle of subatomic matter, the B-meson. At the same time, the archived signals from various reflective, refractive, and diffractive interactions between the emanating x-rays and various material, chemical, and biological samples are recorded in the hope of producing descriptions of the structural arrangement of atoms within those samples.

The laboratory that is the focus of this book began at Cornell University in 1934, or in 1946, or in 1976, or even at some other

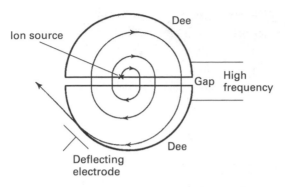

Schematic diagram of a cyclotron. (Bluh and Elder 1955)

time or place, for that is part of the question: Just what *is* the laboratory? Certainly the first cyclotron built in the United States after the one in Ernest Lawrence's laboratory at the University of California at Berkeley was built at Cornell University in 1934 (Heilbron and Seidel 1989).

A cyclotron works by accelerating electrons in an ever-widening spiral inside flat, semicircular copper "electrodes" and then ejecting those electrons at the outer edge of one of the electrodes. By this means, a cyclotron can accelerate electrons to speeds within a few percentage points of the speed of light. Invented during the race to be the first to "split" an atom's nucleus, cyclotrons were important during the 1930s, when subatomic interactions were first explored (Cathcart 2004). In the 1930s, cyclotrons were also used to induce radioactivity in elemental samples, which made them useful in medicine and in biological research. This connection was important for the funding of early cyclotrons, and it enabled continued work in accelerator physics in the United States during the Great Depression, when money for physics was scarce. In the early years, in order to make sufficient quantities of

Milton Stanley Livingston with the 1929 cyclotron in Berkeley (Heilbron and Seidel 1989)

isotopes to meet the medical demand, scheduled crews of physicists worked around the clock at Lawrence's lab, producing prespecified radioactive sources. While this work came to be seen as "downright drudgery," and Lawrence himself became concerned about the bureaucratic transformation engendered by the emphasis on biomedical isotope production, biological work remained a mainstay of Lawrence's laboratory through the years leading up to World War II (ibid.: 306). Lawrence even appeared on radio shows with radioactive salt produced by his machine to demonstrate, by holding the salt on one side of his body and a Geiger counter on the other, the power of the radiation (Heilbron and Seidel 1989: 191). Nearly all the cyclotrons built subsequently during this time in the United States were dedicated to biomedical work, which was still "paramount in acquiring financial support for their construction" (ibid.: 301). By 1940, this meant 20 out of 23 facilities. The Cornell machine initiated this first wave of expansion, although it was one of the few machines built for physics rather than biology. In the years leading up to the war, important contributions to nuclear physics and improvements in the design of cyclotrons were seen to arise from work done with the Cornell machine (ibid.: 301).

After the dramatic success of physics and physicists in the war, particle physics became a national priority. At many universities, alumni of the Manhattan Project and of the Los Alamos Laboratory were granted resources with which to start or to greatly expand programs in high-energy physics. These programs were built on the promise of a new and more powerful generation of circular accelerators: synchrotrons. Rather than accelerate electrons inside a copper plate and then eject them, a synchrotron accelerates electrons in a vacuum inside a circular metal pipe. This makes it possible to accelerate the electrons to about 99 percent of the speed of light. The builders of the 1934 Cornell cyclotron, who had since been involved in the Manhattan Project and the Los Alamos Laboratory, approached Cornell University's president, Edmund Ezra Day, and informed him that they all had offers to go elsewhere to conduct nuclear research. They made it clear that they preferred to stay, but that Cornell would have to provide "adequate opportunities" for "constructive work," including the construction of a synchrotron. In response to this request, Cornell's board of trustees promptly allocated $1.2 million for "specific support of the nuclear studies project" (Schweber 1992: 177). "No decision of the Board of Trustees during my term of office," President Day told the physicists, "has been of greater importance to this institution and to the prospects of scientific research in this country." (ibid.: 178) With additional funds from the Office of Naval Research, the building of an accelerator at Cornell was under way.

In a repeating pattern, the machine operator injects first electrons and then positrons into the storage ring. After each injection, an "experimental run" officially begins. Once the run begins, the operator will continue to make adjustments so that these particles will circulate efficiently in the machine and will collide at just the right place. How much adjustment is appropriate—and even what such adjustment actually is—is a topic of contestation at the lab. For now, over the lab-wide intercom, the machine operator announces to both the x-ray group and the particle group that the run has started. Each experimental run, during which the particle physics group and the x-ray lab conduct their experiments simultaneously, lasts an hour. At the end of the hour, the machine operator announces the end of the run, "dumps" the remaining particles from the ring, and begins a new run.

The Cornell particle physicists began their program of synchrotron building after World War II with a 300-MeV (mega-electron-volt) machine. An electron-volt is the amount of energy gained by a single electron when it is accelerated through an electrostatic potential of one volt. How many electron-volts a synchrotron is capable of producing is directly related to how fast the electrons can travel in that synchrotron. The members of the solid-state physics group at Cornell were aware that a new synchrotron was being built in their midst and that synchrotrons could produce very powerful x-rays. In 1946, when such "synchrotron x-rays" were first noticed by a technician working on a synchrotron at

Schematic of the Cornell 300-MeV synchrotron. The first-ever synchrotron beamline was, on occasion, connected to this device. A swath of x-rays is shown as emanating from the device. (Patterson 2002)

General Electric's laboratory in Schenectady, a member of the Cornell solid-state group immediately visited the GE lab to witness the phenomenon (Hartman 1988: 4). Upon his return to Cornell, members of the solid-state group began to formulate (with help from members of the particle physics group) plans for how they might use the synchrotron on their campus for research. One of the solid-state researchers described his first cautious approach to the particle physics group about the use of x-rays from the synchrotron as follows:

I went one day down [the hall] to see the x-ray physicist . . . about the possibility of using [x-rays] of a low voltage, high current x-ray tube. He was not very encouraging but suggested instead that the synchrotron be used. . . . As a source for my research, the prospect was not the most appealing; the high energy people had their own program, it was a complicated source, and to have two differently oriented laboratories in the same space seemed pretty unrealistic. But the prospect of doing something in my energy range, yet of real interest to the high energy people seemed exciting and useful. (Hartman 1988: 4)

With a colleague, this researcher then approached the head of the atomic physics group about getting some "machine time." The director, it turned out, was "more than agreeable," and they "made plans" (ibid.: 5). Thus, in 1946, the first synchrotron radiation beam time was arranged. The solid-state physicists described the first time they were allowed to use this "machine" as "a satisfying night for us, if not for the high energy crew running the machine, to whom we were indebted" (ibid.: 7). So began a unique collaboration in the world of physics. Describing the initial relationship between the two groups, one of the solid-state physicists involved said: "It was clear that the radiation was there, it was intense, and could be useful. But it was also clear that we were all going to be pirates and that the high energy [particle physics] people were not building their machine(s) with us in mind." (ibid.: 7) For about 10 years, while the 300-MeV machine continued to contribute to particle physics research, the "pirates" conducted important early characterizations of x-ray synchrotron radiation when they could access the machine. Over the next 50 years, this relationship changed. The "pirates" permanently boarded the vessel. Moreover, these pirates had other pirates in their midst.

During laboratory running periods, another pattern of operation repeats. On four or five days of every week the lab conducts experimental runs. Members of the lab's staff and members of the experimental

groups "pull shifts" around the clock. On the other two days, the par-
ticle group conducts "machine studies"—experimental tests that are
used to gather new information about the capabilities of the machine,
the detector, and the x-ray lab. "Machine studies" time is also used to
test new components of the machine, the detector, and the x-ray lab.
The relationship between machine studies and experimental running
is contested at different times in different ways at the lab. Nevertheless,
between these two functions (experimental runs and machine studies)
the laboratory is staffed, open, and operating 24 hours a day, 7 days a
week, 52 weeks a year, although this pattern is broken when the labora-
tory interrupts the running periods, sometimes for months at a time, to
upgrade the facility.

In the mid 1960s, Cornell's particle physics group built a second
and larger synchrotron. When it was commissioned, this synchrotron boasted the highest electron energy in the world: 10
GeV. For more than 10 years it supported a successful experimental program in particle physics, including the exploration of a
newly discovered subatomic particle: the B-meson. This machine
ran through 1975, when the particle physics group submitted a
proposal to the National Science Foundation to "modify" its synchrotron facility. This proposal called for the construction of
an electron-positron colliding-beam storage ring, for which the
existing synchrotron would serve as the injection device. In this
storage ring, electrons and positrons would be accelerated around
at the same time in the same pipe but in opposite directions. (A
positron reacts as a positively charged electron to electric and
magnetic fields.) The particles would weave in and out of one
another in bunches, and would be made to collide, head on, at a
particular point in the ring, thus greatly increasing the energy of
collision. Before this proposal, all colliding-beam machines in the

The tunnel in the 1970s before the storage ring was built. This photo hung in a hallway of the laboratory. The device in the tunnel is the 10-GeV synchrotron

United States operated at the same energy. This proposed storage ring would operate at a different and somewhat variable energy and thus would "substantially increase the flexibility of the national program" (CESR 1977: 24). The phenomena to be studied included possible new hadronic and leptonic degrees of freedom, the spectroscopy of hadronic resonances, hadronic dynamics as revealed by multiple production, photon-photon collisions, the properties of quantum electrodynamics at very short distances, and the relation between weak and electromagnetic processes (ibid.: 9–22).

The proposal was initially reviewed by the National Science Foundation's High Energy Physics Advisory Panel's sub-panel on New Facilities. Although the sub-panel commented favorably "with regard to the technical aspects and physics capabilities," the facility was recommended for construction "only under conditions of a high level of federal funding" (ibid.: 1). Such a high level of federal funding, however, did not emerge immediately. Despite that initial setback, the Cornell physicists, confident in their proposal and sure of their direction, proposed in December of 1975 that they use $450,000 of their already allocated operating budget to begin preliminary work on designs that could be incorporated into the new facility. This request was approved. A similar agreement was reached to use another $1.2 million from the operating budget in 1976 for colliding-beam research and development (ibid.: 2). In 1977, the original proposal was again submitted, requesting $20 million over three years for the design and commissioning of an electron-positron storage-ring facility.

As the storage-ring proposal was again being considered by the National Science Foundation, the solid-state researchers at Cornell began building support in their field for the proposed ring, but for different purposes. By that time, the importance of synchrotron radiation science was internationally recognized. The U.S. Department of Energy was arranging for the construction of two synchrotron rings that would be dedicated radiation facilities. In this general climate, and with the specific proposal for a ring on their campus, the solid-state group at Cornell arranged for a "Workshop on the Application of Synchrotron Radiation to X-ray Diffraction Problems in Materials Science" to be held in the summer of 1977 (Batterman 1977). This workshop raised the solid-state group's institutional visibility and importance further and provided a unified argument explaining why sharing the storage-ring resource and collaboration with the particle group would lead to good science by advancing a variety of research fields. In diffraction physics, alloys and ceramics could be studied.

Surface studies and crystallography of very small crystals could be pursued. Studies of biophysical materials (e.g., muscle and membrane) showed promise. The report of the meeting asserts that synchrotron radiation had "great value in the characterization of materials for industrial processes, particularly in the semiconductor and electronic device industries," and that x-ray techniques to determine "nearest neighbor environments" of electrons in an atom were a valuable resource in "many areas of chemical, solid state, biological, and materials sciences." The report also discussed the use of synchrotron x-rays for "macromolecular crystallography," the determination of the structures of viruses and proteins (ibid.: 5–11).

In September of 1977, the organizers of the workshop submitted to the National Science Foundation a request for $1.4 million over three years. The proposal was to incorporate x-ray beamlines into the new storage-ring facility that was being proposed by the particle group at Cornell (CHESS 1977). This proposal was submitted separately from the particle group's proposal, but it referred to ongoing communication between the groups to show that this request had the support of the particle group. The scientific argument in the proposal followed along the lines of the conclusions of the workshop. In addition to this scientific argument, the proposal outlined the history of cooperation between the high-energy physics group and the x-ray physics group at Cornell. It also noted the timing, the cost effectiveness, and the scientific uniqueness of the proposed synchrotron radiation facility. Specifically, the proposal relied on the following lines of reasoning: Fist, the new facility could provide synchrotron x-ray radiation, which was currently available only at one facility in the United States—a facility that was far from the site of the proposed new lab (ibid.: 5). Second, even though a much larger ring dedicated

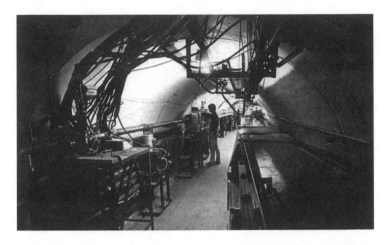

Tunnel with storage ring and synchrotron, circa 1983. (Batterman 1986)

to synchrotron x-radiation was under construction not far away, the proposed lab would come on line more than a year ahead of the dedicated facility, thus providing interim "relief for the ever increasing demands of synchrotron radiation" (ibid.: 7). Third, since the x-ray lab would be parasitic to an already planned ring, the x-ray facility would be extremely cost effective ($460,000 per year) (ibid.: 7). Finally, because of the nature of the storage ring, the proposed radiation laboratory would have available certain high-energy x-rays at intensities that even the national facility would not be able to provide. This would allow the laboratory to "attack unexplored problems of a fundamental nature" (ibid.: 5).

Even though the original colliding-beam ring proposals by the particle physics group in 1975 and 1976 did not mention the possibility of supporting a synchrotron radiation facility, in 1977 the particle physics group submitted an updated design report, noting that "the utility of synchrotron radiation in the physical and biological sciences is well documented" (CESR DR 1977: 11.1) and

Inside the tunnel, circa 1983. (Cornell 2000)

that "while many (synchrotron x-ray) needs can be met by dedicated facilities [the proposed storage ring] can make an important
contribution as an alternative source in the hard X-ray, X-ray, and,
to a lesser extent in the VUV region" (ibid.: 12.1) In the end, the
balance of opportunity cost, scientific possibilities, location, and
the history and spirit of cooperation between these two groups of
physics researchers proved compelling. In 1978 the proposals for
the Cornell Electron Storage Ring (CESR) and the associated synchrotron x-ray facility, called the Cornell High Energy Synchrotron Source (CHESS), were both approved by the National Science
Foundation. With government money and with infrastructure
support from the university, a new kind of laboratory—a hybrid
particle physics-synchrotron x-ray laboratory—was born.

A section of the tunnel in 2001, showing the synchrotron as it bypassed the particle physics detector. (P. Doing)

As each experimental run proceeds, the particle group and the x-ray group are engaged in different kinds of actions, with different rhythms and different concerns. In the particle group's "counting room," above the machine, people (mostly graduate students) watch over a bank of computer monitors as raw data from the electron-positron collisions appear on screens. A scientific result is not immediately discernible, but one emerges after signals collected over several months are stored and analyzed. The counting room makes sure that the collection equipment is working properly, that there are sufficiently strong signals, and that the data are being archived. A few minutes of fluctuation or down time do not matter as long as the machine and the data-collection equipment are working generally. On the floor below, however, just outside the ring area, experimenters and equipment operators watch the intensity and the position of the particle beams closely, since small fluctuations in the

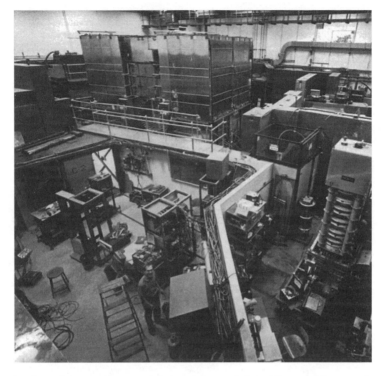

Looking down onto part of the experimental floor. The particle group's detector is inside the large assembly visible in the background. The C-2 x-ray experimental station is visible to the left. (Batterman 1986)

trajectory of the beams, and x-rays that emanate from them, have large effects on experiments that are lined up to detect "scattered" x-rays at extremely precise angles. Most experimenters at the x-ray lab have arranged "beam time" for one or two weeks and probably will not be able to return for a year or more. For many of them, these "run weeks" will determine the topic of a dissertation or a year's publication output. Tensions can run high. The pressure can be felt on the floor. To these groups, delays of even minutes may matter.

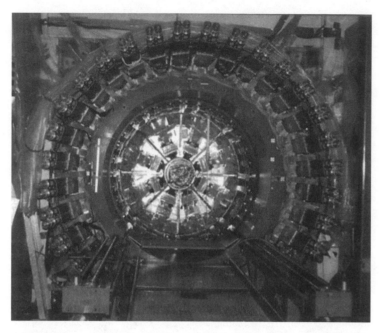

Looking into the particle physics group's detector known as CLEO. (Cornell 2000)

When the new laboratory was launched, the relationship between the x-ray group and the particle physics group with regard to the synchrotron was explicitly stipulated by the particle physics group in memos to the x-ray physicists and to the National Science Foundation. The particle physics group's updated 1977 technical report to the NSF asserts that, although "every attempt will be made to maintain a compatible and productive synchrotron radiation facility," the particle physics group will not lose sight of its mission. "As the storage ring develops," the report states, "high energy physics goals must be preserved" (CESR DR 1977: 12-3). The relationship is further clarified in a memo from the director of the synchrotron lab to the leaders of the x-ray radiation lab dated October 3, 1977. This memo was also attached to the technical report sent to the NSF. After conveying his "enthusiastic support" for the x-ray facility and noting that "the effective use of the unique synchrotron radiation of the storage ring for research purposes is an exciting objective," the particle group's director states "we should assist you in establishing this program in every way that is consistent with our available resources and with the priority of our program in high energy particle physics" and adds these comments:

It is recognized that the primary mission of the CESR laboratory is the high-energy particle physics program. The activities of the CHESS laboratory and the services which the Laboratory of Nuclear Studies will supply must be consistent with this priority. As a consequence, it is expected that CHESS will operate predominantly in a strictly parasitic mode. However, we realize that under special circumstances, it may be extraordinarily useful for CHESS to be able to control the characteristics of the beam and the mode of operation of the storage ring. As a consequence, we agree to make provision for a certain amount of time when the storage ring will operate in the mode specified by the CHESS Laboratory directorate. (CESR DR 1977, appendix A)

For 3 percent of the time, the x-ray facility would control the ring. For the other 97 percent of the time, x-ray operation would be "strictly parasitic." From pirates to parasites, the solid-state physicists had come a long way. Even if constrained, they now had official status and access to a unique and powerful x-ray source. With this pact in hand, both groups got down to the business of vigorously pursuing the new scientific opportunities that lay before them.

For several decades, the laboratory thrived. In a scientific landscape in which the colossal Superconducting Supercollider had been approved and was under construction and dozens of dedicated x-ray synchrotron laboratories were coming on line around the world, the particle group at the Cornell laboratory contributed importantly to B-meson physics, and the solid-state group a wide variety of research in materials science and biology. The laboratory's staff grew to several hundred, housed in offices located above the synchrotron and the storage-ring tunnel, the x-ray experimental floor, and the adjacent machine shop and fabrication space.

Beginning in the 1990s, however, an important change occurred at the lab and in the field. Protein crystallography, once only a small part of the x-ray research (which was itself secondary to the particle physics work at the lab), gained more and more prominence at the lab and in science at large. As anticipation grew that the Human Genome Project would result in full control over disease and health, and as developments in the combination of molecular biology and information technology advanced, funding in science flowed toward research with biological applications. Meanwhile, public interest in particle physics waned, and funding for such research was scaled back. The Superconducting Supercollider project was halted in the midst of construction (Kevles 1995, 1997). As scientific prestige in the form of publications, funding, and professorships accrued to research in biology in general and in the structure of viruses and proteins in particular, including the awarding of the Nobel Prize in chemistry in 2003 for protein crystallography work done in part at the Cornell lab, practice at the laboratory changed. Who could properly produce technical and scientific knowledge, and how, was different at the end of my study than it had been at the beginning. In this book, I explore these changes at the lab and wrestle with their implications.

The wind gusted through the glass door as I entered the laboratory building from the parking lot across from the soccer fields. The lobby area was nondescript—just a small coffee table and a chair of metal and fabric beside an elevator—except for one item. Leaning forward from the side wall of the high-ceilinged foyer was a life-size praying "maiden" carved from wood and taken off of the prow of a sailing ship. With her hands folded in front of her, she watched over those who passed in and out of the lab, silently guiding their passage and the journey of the machine. As I was envisioning her earlier life at the front of her previous vessel, pressing through the waves of past travels, the elevator's bell chimed and its doors slid open in front of me. I stepped forward and turned around as the doors glided shut. My stomach lurched as the elevator car bumped into motion.

2 Practice and Product: Analyzing Science

In its most basic formulation, the central question that drove my exploration of the lab—a question that has divided studies of science since Thomas Kuhn's book *The Structure of Scientific Revolutions*—is whether the product of scientific practice, the very content of science, is contingent on that practice. In other words, does the natural world exist outside of efforts to describe it? (A good explanation of this division can be found in Pinch 1997.) The answer to this question, of course, has enormous implications for the world we live in, for governance, for ethics, and for humanistic inquiry in general. Beginning in the late 1970s, researchers of a philosophical bent sought to settle this question by leaving the philosopher's armchair and entering working laboratories with the express mission of exploring how the dynamic and changing world of contingent real-time scientific practice might be implicated in the status of the products of the scientific enterprise. News of their successes spread rapidly. In northern California, close scrutiny of laboratory-bench conversations and "shop talk" showed how what was "seen" as the output of instrumentation was guided by intricately choreographed vernacular in the course of experimentation (Knorr Cetina 1981; Lynch 1985). In San Diego, the laboratory of the eminent Jonas Salk was engaged, and scientists were seen to be producing deconstructable "inscriptions" rather than ordinary facts. (Latour and Woolgar 1979). In Britain and deep under America's Great Plains, scientists looking for gravity waves and solar neutrinos were seen to be relying on their "enculturation" in the sociality of scientific practice to generate knowledge of the natural world (Collins 1985; Pinch 1986). All these new studies were lauded for their attention to the details of laboratory life, and their daring assertions took hold both in the arena of scholarly inquiry and in that of policy making. Over the past several decades, this work has influenced the fields of anthropology, philosophy, sociology, and literary studies, has been taken up by policy analysts, and has been referred to in regulatory and legal decision making, all the while contributing to the growth and influence of the field of Science and Technology Studies. Asserting that science is fundamentally contingent on practice, these "lab studies" formed a corpus of intellectual work that has had provocative and profound implications for the project of intellectual inquiry and for the essence of political citizenship. To explain the project of this book and my motivation for writing it, a question must be asked of these pioneering studies: Just how did they accomplish what they claimed, and were purported, to have accomplished? How did they, as Karin Knorr Cetina put it (1995: 141), disclose "the process of knowledge production as 'constructive' rather than descriptive" and show the "'made' and accomplished character of technical effects"?

The Shop Floor of Facts

In *Art and Artifact in Laboratory Science: A Study of Shop Work and Shop Talk in a Research Laboratory*, Michael Lynch explains that the "science that exists in practice is not at all like the science we read about in textbooks" in that "successful experimentation would be

impossible without . . . decisions to proceed in ways not defined
a priori by canons of proper experimental procedure," and that
"a principled demarcation between science and common sense
no longer seems tenable" (1985: xiv). Lynch points to how fluid
judgments of sameness and difference, conversational accounts,
vernacular in the laboratory, practical limitations, and negotiations—the processes of scientific practice—play into the acceptance and rejection of reality on the laboratory floor. Lynch's
descriptions of laboratory life are quite compelling. In real time,
researchers struggle to negotiate what is "understood" in the
moment so that a subsequent action is justified. The descriptions
of the myriad of micro-social assertions and resistances put to
work in the lab is rich, and that such negotiations are part and
parcel of moment-to-moment practice rings true to anyone who
has participated in scientific practice—it can be a mess of piece-
meal and ad hoc steps and justifications. But how does Lynch link
the contingent world of the laboratory to the status of any particular enduring fact that the laboratory is seen to have produced?
In view of Lynch's introductory statements, one would expect an
enduring fact to be subjected to his analysis and his method in his
book. However, this project is not taken up directly. Indeed, at the
end of the book Lynch writes that "whether agreements in shop
talk achieve an extended relevance by being pre-supposed in the
further talk and conduct of members or whether they are treated
as episodic concessions to the particular scene which later have
no such relevance, cannot be definitively addressed in this study"
(ibid.: 256). He then asserts that "the possibility that a study of
science might attain to an essentializing grasp of the inquiry
studied is no more than a conjecture in the present study" (ibid.:
293). Lynch's study, then, is not a direct challenge to the "principled demarcation" of science; it merely implies that a challenge
could be brought. In *Art and Artifact*, we are invited to consider

the possibility that the detailed and compelling dynamics of day-to-day laboratory work presented might have implications for demarcating the products of science from the contingency of laboratory life, but by Lynch's own explicit acknowledgement we are not presented with an account of how this is so for a particular fact claim—that is, how any particular episodic agreement, as a matter of practice, became a fact with "extended relevance."

Where Lynch left off, though, others pressed on, explicitly pushing their analyses to wrestle with the status of particular fact claims.

Indexical Manufacturing

In *The Manufacture of Knowledge: An Essay on the Constructivist and Contextual Nature of Science*, Karin Knorr Cetina takes up the challenge to ethnographically demonstrate the local construction of an epistemically demarcated fact. Explaining her project, Knorr Cetina writes:

In recent years, the notion of situation and the idea of context dependency has gained its greatest prominence in some microsociological approaches, where it stands for what ethnomethodologists have called the "indexicality" of social action. . . . Within ethnomethodology, indexicality refers to the location of utterances in a context of time, space, and eventually, of tacit rules. In contrast to a correspondence theory of meaning, meanings are held to be "situationally determined," dependent only on the concrete context in which they appear in the sense that "they unfold only within an unending sequence of practical actions" through the participants' interactional activities. (1981: 33)

The shop floor of the lab, again, is the place to find this situational world of practical action, and indeed Knorr Cetina finds it. Like Lynch, she provides compelling ingredients for a sociopolitical analysis of the technical. She astutely observes the subtle way in which power is "played out" between scientists for access and control of resources and authorship and credit (ibid.: 44–47), and she argues convincingly that a series of "translations" from one context to another is the mill from which new "ideas" are generated and pursued in the course of laboratory research (ibid.: 52–62). She further asserts that larger "trans-scientific" fields are ever present in the day-to-day activities and decisions of laboratory researchers (ibid.: 81–91). Moreover, she goes further than Lynch in pursuit of a political account of a technical fact as she follows a particular technical fact through to its culmination in a scientific publication. Knorr Cetina points out that the active, situated work on the part of researchers as they negotiate the contingent, messy life-world of the laboratory cannot be found in the final official published description of the episode, which reads like a high school textbook's account of the scientific method, with its orderly sequence of hypothesis, experiment, and results. The question, again, is: How, precisely, does the fact that this work took place and was subsequently erased relate to the status of the particular technical fact claimed by the scientists in their publication on that subject? Precisely how is the technical claim presented by practitioners that "laboratory experiments showed that $FeCl_3$ compared favorably with HCl/heat treatment at pH 2–4 with respect to the amount of coagulable protein recovered from the protein water" (ibid.: 122) implicated as "situationally determined"?

Despite its ambitions, Knorr Cetina's study is like Lynch's in that it does not actually confront the picture of science as a demarcatable realm of knowledge head on, but instead sidesteps the distinction between contexts of discovery and proof. All scientific papers erase contingency, but not all of them "produce" facts. It isn't the erasing in and of itself that coerces, or doesn't coerce, the acceptance of a fact claim. Knorr Cetina does not address why *this* erasing worked in *this* situation while other erasings do or have not, and that is the crux of the matter for a study that seeks to assert that knowledge production is "constructive" rather than "descriptive."

Where Knorr Cetina leaves off, however, Bruno Latour and Steve Woolgar press on in spectacular fashion.

Contingent Inscriptions

In their 1979 study of Jonas Salk's laboratory at the University of California at San Diego, *Laboratory Life: The Social Construction of Scientific Facts* (later re-titled *Laboratory Life: The Construction of Scientific Facts*), Latour and Woolgar explicitly set out to show how the hardest facts—scientific facts—could be deconstructed. They assert that "a close inspection of laboratory life provides a useful means of tackling problems usually taken up by epistemologists" (1979: 183). To make their point demonstrably, Latour and Woolgar focus not on a small fact but rather on one that resulted in Nobel Prizes and historical prestige for a legendary laboratory: the discovery at the Salk Institute that thyrotropin-releasing factor (or hormone), TRF (or TRH) is, in fact, the molecular compound (in shorthand) Pyro-Glu-His-Pro-NH_2. This discovery was important because it provided a causal explanation for the effects and actions of the hormone. In considering Latour and Woolgar's account of this fact claim, we must ask: Exactly where are the points at which their account of the "discovery" of TRF(H) as Pyro-Glu-His-Pro-NH_2 implicates contingent local practice in this enduring, accepted fact? Two critical points in the "TRF(H) as Pyro-Glu-His-Pro-NH_2" story bear close scrutiny in this regard. The first is the point at which the acceptable criteria for what counted as a

statement of fact regarding TRF(H) changed among the practitioners. Whereas isolating the compound in question had been seen
as impracticable, and therefore as irrelevant for making statements
of fact about TRF(H) (owing to the fact that literally millions of
hypothalami would have to be processed, and that would mean
the killing of too many laboratory test animals and would take too
much time and resources), there later came a point at which the
field decided that such a "big science" project was the only way
to obtain acceptable evidence of the actual structure of TRF(H).
Old claims about TRF(H) were now "unacceptable because somebody else entered the field, redefined the subspecialty in terms of
a new set of rules, had decided to obtain the structure at all costs,
and had been prepared to devote the energy of 'a steam roller' to
its solution" (ibid.: 120). The success of this intervention, according to Latour and Woolgar, "completely reshaped the professional
practice of the subfield" (ibid.: 119). This, then, is an episode ripe
for anti-demarcationist explanation. The criteria for judging facts
changed as a result of local, contingent, and historical actions! A
particular person entered the field and changed the rules of the
fact-finding process. Now the move would be to explore why
and how this happened and was sustained—why it worked and
endured. According to Latour and Woolgar, the reason this pursuit succeeded as valid, proper science, and as the new touchstone
of fact claims about TRF(H), rather than being seen as golem-like
excess and unnecessary waste, is as follows:

The decision to drastically change the rules of the subfield appears to have
involved the kind of asceticism associated with strategies of not spending
a penny before earning a million. There was this kind of asceticism in the
decision to resist simplifying the research question, to accumulate a new
technology, to start bioassays from scratch, and firmly to reject any previous claims. In the main, the constraints on what was acceptable were

determined by the imperatives of the research goals, that is, to obtain the structure *at any cost*. Previously, it had been possible to embark on physiological research with a semi-purified fraction because the research objective was to obtain the physiological effect. When attempting to determine the structure, however, researchers needed absolutely to rely on their bioassays. The new constraints on work were thus defined by the new research goal and by the means through which structures could be determined. (1979: 124)

That a value of "asceticism" would pervade scientific practice and would provide the basis for acceptance of a new way of making facts is the assertion. But why would asceticism be the driving force that coerces fact demarcation in this situation but not in others? Here we are left at the same point as with Knorr Cetina and Lynch.

At another point in Latour and Woolgar's account, however, local practice is specifically, and more compellingly, implicated in the subsequently "produced" fact. At the end of the account of the emergence of TRF(H), Latour and Woolgar describe another crucial turn in the making of the fact as a fact: contestations over decisions about the sameness or difference of various curves obtained with a chromatograph. Since the nature of TRF(H) rested on judgments of sameness and difference for the curves made with this device, and such judgments were (and could always be in principle, according to the authors) challenged, Latour and Woolgar assert that the status of the structure of TRF(H) was in epistemological limbo. How was this episode closed off so that its product could endure as a scientific fact? At this point, Latour and Woolgar describe how a device from physics—the mass spectrometer—settled the issue. They tell us that the scientists "considered that only mass spectrometry could provide a fully satisfying answer to the problem of evaluating the differences between natural and synthetic (a compound made to be like) TRF(H)," and that "once a spectrometer had been provided, no one would argue anymore" (1979: 124). Here, it would seem, is the critical juncture for the usual picture of science as demarcatable from contingent practice to be challenged. If the mass spectrometer settled the matter, what is contingent about the mass spectrometer? At this point, after we have followed the journey of TRF(H) all this way, Latour and Woolgar inform us that "it is not our purpose here to study the social history of mass spectrometry." Well, if mass spectrometry did in fact decide the matter such that the compound now exists as a matter of fact rather than a contestable assertion, it should have been Latour and Woolgar's *main* purpose to analyze mass spectrometry as a "social historical" phenomenon. Latour and Woolgar then state that the nature of TRF(H) will "remain unambiguous as long as the analytical chemistry and the physics of mass spectrometry remain unaltered" (ibid.: 148). There is no clear route from the contingent world of the shop floor to the enduring fact of TRF(H) Pyro-Glu-His-Pro-NH$_2$ other than via the inference that, in principle, a thoroughgoing deconstruction along those lines could be undertaken. Again, that deconstruction has not been done for us.

The issue is the relation between contingent, local practice and the status of enduring trans-local, trans-temporal technical facts. In a world of contingent practice, what establishes that a particular fact endures as a fact? In the early laboratory studies, the only time the endurance of a particular fact is specifically addressed is when Latour and Woolgar first meekly gesture to asceticism to explain how the accepted criteria for a fact claim changed, and then settle on the atomic mass spectrometer to account for how the TRF(H) controversy was eventually decided and made to endure.

Enculturated Assertions

In his 1985 book *Changing Order: Replication and Induction in Scientific Practice*, Harry Collins asserts that there is a fundamental regress in experimental replication. If a new phenomenon is purported to be discoverable by experiment, and an experiment is constructed to do so yet does not, there is no way, in principle, to determine whether the fault lies with the experiment or with the assertion of the undiscovered phenomenon, because the phenomenon may, after all, not be discoverable! Conversely, if the experiment "registers" the phenomenon, it could be an effect of the instrumentation. Animated by this principle, Collins looks to a specific scientific controversy involving early experiments on gravity waves in order to explain how this dilemma is dealt with in the actual practice of doing science. One of the scientists in Collins's study had been making a claim for the detection of gravity waves that went against the prevailing theory of gravity waves and also against the results from other experiments designed to detect the waves. The scientists claimed that his experiment had detected a much higher flux of gravity waves than theory had predicted, and so the theory must be wrong, and other detectors that backed up the theory must have been improperly made. When an "electrostatic calibrator" was brought in to simulate gravity-wave input for the scientist's detector in an effort to overcome the objections of experimenters who supported the prevailing theory, it was found that his detector was 20 times less sensitive than the other detectors. After this, the claims for high-flux gravity waves were dismissed. Collins points out that according to the experimenter's regress, the investigator could have claimed that the electrostatic calibrator did *not* simulate high-flux gravity waves (which could not be simulated, as they had not been discovered) and that the fact that high fluxes were detected with only his particular kind of detector, even though it was less sensitive to the calibrator, gave important information *about the nature of gravity waves*. Well, this is just what the investigator did, only it didn't wash. The investigator's claims in this regard were seen as "pathological and uninteresting" by the other scientists. "The act of electrostatic calibration," Collins explains, "ensured that it was henceforth implausible to treat gravitational forces in an exotic way. They were to be understood as belonging to the class of phenomena which behaved in broadly the same way as the well-understood electrostatic forces. After calibration, freedom of interpretation was limited to pulse profile rather than the quality or nature of the signals." (1985: 105)

Collins assures us that the dismissal of the high-flux claims was not determined by nature. It was *the investigator* who had the agency, who "accepted constraints on his freedom" by "bowing to the pressure" to calibrate electrostatically, and thus "setting" certain assumptions beyond question. Collins asserts that the investigator would have been done better to refuse this constraining electrostatic calibration. But what of this pressure on the investigator to calibrate? What gave it such force that the investigator capitulated? Where did it come from? Who controlled it? Why did it work? When reading Collins's account of gravity-wave experimenters, we find ourselves in a similar situation as with Latour and Woolgar: at the crucial juncture where controversy ends and a fact is born via a differentiating technology, we are left to wonder just how contingent practice coerced the endurance of the particular fact claim. Again, the account reads like a conventional treatment of science—calibration settled the dispute. We are simply told by Collins that in principle the episode could have gone otherwise *and been accepted as scientific*.

It is important to note here that long-running arguments between Latour and Collins about how they each approach explaining fact-making as constructive rather than descriptive are off the mark with respect to the project of this book. The important point to understand is that *both* Collins and Latour and Woolgar go against the admonition asserted by Lynch that a constructivist argument must not to be preoccupied with "defining, selecting among, and establishing orders of relevance for the antecedent variables that impinge upon 'actors' in a given setting" (Lynch 1985: xv). In other words, that practice itself must support the enduring legacy of its products. Whether it is the social construction that is claimed to be demonstrated in Collins's or in Latour and Woolgar's (1986) non-modern "construction" of the "actor-network" theory (in which the social and the natural are treated symmetrically, but both as constructive) doesn't matter. Both studies break with the plane of practice in which method is used tautologically, bring in an element or elements from the outside to account for the endurance of the facts under question, then argue over which is the better way to do so (Collins and Yearly 1992). These subsequent arguments have to this day not furthered the project of implicating local practice in the ontological status of any particular scientific fact.

In his 1986 book *Confronting Nature: The Sociology of Solar Neutrino Detection*, Trevor Pinch describes *in situ* the first experimental attempts to detect solar neutrinos, and goes a step beyond Collins in explaining how practice could secure an enduring legacy for a constructive fact. Pinch describes a controversy over the amount of solar neutrinos that are seen to be detected by large vats of chemical detectors located deep underground in abandoned mines. As it was with Collins, the linchpin of closure is calibration. But Pinch goes further than Collins, asserting that the linchpin of calibration is credibility. He then endeavors to explore this "credibility" by examining just how his experimenter was able to negotiate the relationships necessary to ward off critics of his detector. Pinch explains how the experimenter in question (Davis) would give the details of his experiment, which detected higher amounts of solar neutrinos than generally expected, to a group of nuclear astrophysicists—members of the theoretical group by which the accuracy of any assertion about solar neutrinos would be judged. This enabled the astrophysicists to "put their criticisms directly to [Davis]" rather than through the medium of publication. Pinch notes that by the time a criticism did appear in print "the battle had largely been won by Davis" (1986: 173). Pinch also points out that Davis was willing to go through the "ritual" of testing all sorts of "implausible" hypotheses brought forth from the astrophysicists. By taking on all comers, Davis performed "an important ritual function in satisfying the nuclear astrophysicists, and thereby boosting the credibility of his experiment" (ibid.: 174). And Davis, through his informal relationship with the astrophysicists, stayed within the boundaries of his "acknowledged expertise," to credible effect. As Davis himself put it, "this all started out as a kinda joint thing . . . and if you start that way you tend to leave these little boundaries in between. So I stayed away from forcing any strong opinions about solar models and they've never made much comment about the experiment" (ibid.: 173).

In Pinch's account, the nuclear astrophysics group was the touchstone for what counted as a proper experiment, and Pinch investigated the practical matter of the negotiation of relations of authority, such as work with the "little boundaries," which reflexively reinforced the "credibility" used to close off the contingency of a technical fact. In this respect, Pinch pushed beyond Latour and Woolgar and Collins to relate contestation over a specific fact claim to the authority of a plausible criteria-setting entity. But then why did these "little boundaries" work this time when they might not work at other times? How did this group achieve the lasting authority to set the criteria of discovery?

Synergistic Currents

In *Constructing Quarks: A Sociological History of Particle Physics*, Andrew Pickering also takes on the relationship between groups of theorists and experimentalists as grist for the mill for a constructive view of scientific facts. His stated goal is to "interpret the historical development of particle physics, including the patterns of scientific judgments entailed in it, in terms of the dynamics of research practice" (1984: 8). For Pickering, science is a practice in which there is "a symbiosis between natural phenomena and the techniques entailed in their production, wherein each confers legitimacy on the other," and "such a symbiosis is a far cry from the antagonistic idea of experiment as an independent and absolute arbiter of theory" (ibid.: 14). Pickering explicates this dynamic with regard to one particular fact claim: the "discovery" of weak neutral currents. With specific regard to these neutral currents, Pickering writes:

How should the relationship between the discovery of the weak neutral current and the development of unified electroweak gauge theory be conceptualized? In the archetypical "scientist's account," the former would be seen as an unproblematic observation and an independent verification of the latter. But can this view withstand historical scrutiny? I want to suggest that it cannot. . . . Two connected arguments will be involved. First, I will argue that the observation reports which emanated from CERN in the 1960s and 1970s were all grounded in interpretive procedures which were pragmatic and, in principle, questionable. . . . I will then show that there were significant differences between neutrino experimenter's interpretative procedures in the 1960s and those of the Gargamelle collaboration (in the 1970s), and that the differences were central to the existence or non-existence of the neutral current. Finally, I will argue that the communal decision to accept one set of interpretative procedures in the 1960s and another in the 1970s can best be understood in terms of the symbiosis of theoretical and experimental *practice*. (ibid.: 187–188, emphasis added)

Pickering notes that in bubble-chamber experiments in the 1960s it was accepted practice to "filter" the data collected with an "energy cut" that excluded signals beyond a certain upper limit of energy (in this case, 1 GeV) and that left a relatively small number of signals unaccounted for and ascribed to background noise. Pickering points to this "cut" as a pragmatic move. He notes that, when a researcher's calculations of the background signals did not account for all of the "background" signals measured, it was accepted that the calculations were wrong, and not that a new phenomenon had been observed. Pickering then notes that in the 1970s the Gargamelle bubble chamber, which was much bigger than the bubble chambers of the 1960s and which had an "improved" neutrino beam, had a relatively large number of "background" signals left over after the usual 1-GeV cut. Now, on the basis of a background analysis like the one done by the researcher in the 1960s, it was asserted that the "extra" background was indeed due to the weak neutral current. Pickering points out that all the objections that were raised in the 1960s against the calculation of the proper background could indeed have been raised again in the 1970s, but after several months "the discovery (of the weak neutral current) came to be regarded as established" (ibid.: 192). He concludes that "the 1960s order, in which a particular set of interpretive procedures pointed to the non-existence of the neutral current, was displaced in the 1970s by a new order, in which a new set of interpretive procedures made the neutral current manifest." "Each set of procedures," he continues, "was in principle questionable, and yet the HEP community chose to accept first one and then the other." (ibid: 193) Pickering has a reason for why this happened: the symbiosis of experiment and theory. He casts the decision to accept different interpretive procedures as opportunistic:

The neutrino experimenters did reappraise their interpretive procedure in the early 1970s, [but] the new procedures remained pragmatic and were, in principle, as open to question as the earlier ones. But, like the earlier ones, the new procedures were sustained in the 1970s within a symbiosis of theory and experiment. In adopting the new procedures, the neutrino experimenters effectively got something for nothing. They had only to take seriously the kind of neutron background calculations performed (but not taken seriously) in the 1960s in order to bring into being a whole new phenomenon: the neutral current. . . . Murray Gell-Mann noted that "the proposed electroweak models are a bonanza for experimentalists," and so I proved. (ibid.: 194)

Programs of neutrino experimentation grew at major laboratories. The discovery of the weak neutral current was a "bonanza" for theorists too. "By accepting the discovery reports and, implicitly, the neutrino experimenters' new interpretive procedures," Pickering writes, "gauge theorists armed themselves with justification for their contemporary practice and with subject matter for future work." (ibid.: 195) The neutral current was "both the medium and the product of this symbiosis, and acceptance of novel interpretive procedures was the price of its existence" (ibid.: 195).

Like Pinch, Pickering went further than asserting that interpretations could have been otherwise and asserted why they were accepted as so. We must ask, though, whether Pickering's justification for writing off the scientists' later understanding— that actually there were neutral currents in the 1960s, but they weren't recognized as such—is compelling. Pickering asserts that the experimental and theoretical fields are mutual touchstones for interpretations that endure, and each of these fields would have interests that, if served, would continually hold up a particular interpretation. We are still left, however, with only Pickering's assertion that decisions that were made could, in principle, be otherwise and still be science.

This is the question, then: How can the practice of judging scientific facts change and endure in such a way as to continually secure the status of fact claims that emanate from that practice? Of the pioneers, Pinch and Pickering come the closest to giving a compelling answer. But even they only allude to the power of particular groups in determining the ongoing criteria for judging facts; they do not explore just how those groups came to hold such power, or how they manage to enforce their authority over time such that is not subsequently challengeable as science. The project of accounting for the *enduring* legacies of practice was, in fact, abandoned shortly after it was begun in laboratory studies, despite the claims of the authors themselves and the subsequent claims of the field at large to have shown scientific facts to be constructive rather than descriptive. The project of this book is to give a more compelling means of answering the question of how change in scientific practice can endure, and to explore the implication of that change for the status of scientific facts.

Embedded Observation

In pursuing this project, I followed a path that continually brought me back and forth between the working world of the laboratory and a world of philosophical reflection on that work. I started work at the laboratory under study in 1991 as an x-ray laboratory operator. In 1993 I was promoted to assistant manager of operations for the x-ray laboratory. I continued to work at the lab in that capacity until 1999. During that time, I also began pursuing a doctorate in Cornell's newly formed Science and Technology Studies department. Over the course of my 8 years at the laboratory, I participated in a myriad of conversations, experiments, accidents, successes, failures, meetings, and presentations. I heard different and changing opinions of the nature of the laboratory and laboratory practice from lab members who, like me, were continually trying to understand what was going on for their own sake as laboratory members. As one scientist explained it to me, "it's like trying to understand an ant colony, and you are one of the ants" (Field Notes, Book 4, 3/28/93). During that time, I completed my coursework and the examinations for my doctorate.

About a year after I began working at the lab, I began keeping field notes. It was convenient for me that everyone at the lab carried notebooks in which they would keep technical notes for themselves about different projects. It was easy to record notes from conversations with people who I thought were interesting for my longer-term study of practice and change at the lab. Other people also took notes about conversations at meetings and presentations. As time went on, I decided which episodes would be of interest to my study and made it a point to particularly follow those episodes. I left the lab in 1999 to finish my doctorate. In writing the account that would lead to this book, I would read my field notes on the episodes that form the main chapters of this book. As I did so, the quotes, descriptions, and thoughts would trigger memories of my time at the lab. In this book, I have referred to specific quotes of conversations and the marking of specific episodes in my notebooks. When making such quotes, I have noted the notebook number and the specific date of the entry. In filling out the rest of the episodes, I have used the phrases "I remember" and "I recall" to make clear that I am writing from the memory of my experience. The general narrative mode, then, is of me writing about events as I remember them as anchored by specific references to the ethnographic field notes that I took at the time. In filling out the episodes in this study, I have also used excerpts from several different types of laboratory documentation. I have used entries in experimental and operational logbooks and electronic logs to mark the specific dates of particular events. I have used job advertisements and promotional literature released by the laboratory to bring out different kinds of conceptions of different laboratory members and self-conceptions of the laboratory as a whole. I have also used the descriptions of newspaper reporters and other visitors to the lab, and a secret electronic log that was kept by the operators unbeknownst to the scientific staff at the lab, to explore how laboratory members present themselves and their work to particular audiences. I have used materials from intra-lab meeting presentations and from talks at extra-lab scientific conferences to mark specific moments and gestures. I have also used material from funding proposals to the National Science Foundation in analyzing how the laboratory places itself with respect to the larger field of synchrotron science and the constellation of synchrotron laboratories in the United States and in the world.

This method of accounting and presentation, like any method, afforded me both advantages and disadvantages. The primary advantage is that this is the first laboratory study written by a participant in the technical episodes that are the subject of the study. In this regard, my direct involvement with the episodes that were at the time under my philosophical consideration enabled me to push certain lines of analysis with regard to the status of scientific claims *in real time at the lab with the practitioners* and see what the reactions would be. This gives my analysis a recursive dynamic that is unprecedented in laboratory studies. My position as a lab member, however, could have limited my analysis. As a working member of the laboratory, I had to remain in my position to a certain degree. I could not conduct broader sociological surveys of the laboratory members (which might have disclosed interesting and salient aspects of their experiences at the lab) and, to my mind at the time, still be seen as a proper technical member of the lab. Also, there were inevitably modes of authority and power to which I did not have access. Another positioning might have engaged issues of gender or the role that class might play in the working practice of the lab differently than I have done in this book.

In the next three chapters, I describe episodes of change at the lab involving operators, physicists, and biologists. In chapter 3, I start with interactions between operators and scientists at the lab and explore how each group's assertions with regard to the knowledge-producing abilities of the other were simultaneously assertions of workplace control at the lab. In chapter 4, I use this kind of analysis to explore interactions between physicists at the x-ray lab and physicists in the particle physics group over the diagnosis of technical malfunctions and proper operation of the storage ring during the conducting of x-ray and particle physics experiments. In chapter 5, I relate the changing relationship between the x-ray physicists and the biologists at the lab to contestations over the development of experimental instrumentation for protein crystallography and the method of proper experimentation with regard to protein crystallography experiments themselves. In all these episodes, the very meanings of the technical identities I have listed, with their concomitant knowledge-producing abilities, are at stake in laboratory interactions and negotiations. They are outcomes of assertions and resistances of epistemic politics, whereby who can know what, and how, is negotiated and made to endure in practice. In the last chapter, I consider the implications of the rise of protein crystallography in synchrotron science for the status of particular fact claims emanating from the lab.

3 "Lab Hands" and the "Scarlet O": Operators, Scientists, and a Feel for the Equipment

The elevator jolted to a stop and its doors slid open. I was confronted by a wall of lights, numbers, electronic traces, and accompanying hums, clicks, and buzzes. The sounds were chaotic. I lingered in front of the synchrotron control room, the gateway to my new world, before I made my way down the hall to the x-ray laboratory reception area. Interview days have a peculiar energy. Your senses are heightened. A potential world is opening before you, yet you see only glimpses, flashes of the future. You look for clues in the surroundings, the conversations, and peoples' demeanors. My friend had told me about this job. He said it was great. There were lots of chances to do interesting things and lots of smart people. He told me that the receptionist was very nice, and he specifically told me to wear jeans, not slacks. Coming from a year of corporate life, I compromised—black jeans. The interviews were fairly succinct. My questioners held their cards close to their chests. I couldn't tell if they liked me. My master's degree work had been in space plasma physics, not synchrotron radiation. Was that relevant? Had I taken mostly math classes, or had I worked with the radar equipment? I told them about the circuit I had designed and built for a radar receiver. I sensed disappointment when I revealed that I didn't work on my own car. I quickly noted that it was more efficient to have someone who does it all the time do it, and that I was very busy. When prompted, I said of course I did the little things like change my own oil (although in fact I never had). The offices were messy. The hallways were chaotic and cluttered. I didn't know aluminum foil could be put to so many uses.

I was walked past a voluptuous Georgia O'Keefe print hanging on the wall and stopped at a kind of portal surrounded by lights and signs (e.g., "Radiation Badge Must Be Worn Beyond This Point," "Synchrotron On"). With some trepidation, I clipped on a radiation badge. Why did I need it? What was the danger? If there was no danger, why have a badge? Did the badge protect me somehow? My guide politely answered my questions as we made our way through the entrance to the experimental floor. The badge was a recording device, required by law. No person in the history of the lab had ever recorded an unhealthy dose of radiation. In fact, the ambient radiation underground was less than one usually encounters in the course of a day. I was feeling a bit sheepish for asking such questions when we walked into a cavernous hall whose ceiling must have been 50 feet high. I was dwarfed by giant gritty lead slabs, huge chrome tanks, and gleaming steel pipes. The doors of experimental rooms slid open and shut. Alarms beeped and then were quiet. A gentle bell chimed continually. A large electronic billboard near the top of the wall at the end of the "floor" indicated that the magnets were powered up and that the storage ring was operating. I looked up and scanned across the ceiling to the far side of the hall. Just below a gouge in the concrete 40 feet above the floor was a sign that read "Top quark went through here." I didn't get the joke. My tour guide said that he loved the sign.

In the interviews, they had told me that the job of an operator here was a good way to see and participate in front-line research, but that certain tasks and procedures had to be done to keep the lab running, and the operator was responsible for these. Well, for me the point of working here was to have a chance to get into some real science, to be where the action was. What were a few tasks? I would be called an x-ray lab operator. Well, I didn't much like the sound of that—as if I "operated" a piece of machinery? I had a Bachelor of Science degree in electrical engineering from an Ivy League school, and I was one project away from a Master's. I had already worked as for a corporation as a design engineer. Now I was some kind of operator of something. "Oh well," I thought, "what's in a name?" The main thing was that I could taste the science, the research, the action.

Well, there is a bit more to the story. I had also begun an introduction to something else. I had started taking some classes on the campus above the synchrotron in the university's Science and Technology Studies department, and I was becoming interested in various ideas about the nature of science. What is science? Where do you find it? How do you know when it is happening? In my classes, I read work in the sociology of scientific knowledge that took the view that the supposedly straightforward scientific method of formulating hypotheses and subjecting them to tests wasn't so straightforward after all. I was introduced to assertions that Karl Popper's theory of falsification (1963) and Robert Merton's description of norms in science (1957) did not fully account for how science really works. According to this new work, the way to find out about science was to go into the laboratory and experience science in all its messy, confusing, complicated glory. I read of laboratory rituals and shop talk, of taken-for-granted instruments and procedures, and of inarticulable tacit knowledge and skills (Lynch 1985: 3). Some of the ethnographers of laboratories held degrees in physics but had switched over to the humanities or the social sciences. They said that their understanding of the

content of physics was what distinguished them from the previous generation of sociologists, and that sociology of science
wasn't just about how science was organized or funded but also
about what science produced. Sociology of science needed a
sociology of knowledge to go with it. I went for this job at the
lab in part because I had a notion that I would somehow have
a chance to study science itself. I remember the long pause that
ensued when I told the professor of a class I had taken called
"Inside Technology" that I was going "do one of those laboratory studies like we heard about in class." He was very polite. He
gave me a list of books to read, and suggested I attend his upcom-
ing class in "Qualitative Methods for Studying Science." I knew
that something could happen if I could just get into the mix,
get involved with some science. If there was more to the production of technical knowledge than how it is presented afterward in journals, talks, tours, and lectures, I was going to go and
find it!

Learning by Doing: An Operator's Initiation

On one of my first days, I was brought to the central control area
of the lab, where several operators were waiting to begin their day.
I went with one of the operators to change a tank of compressed
helium. I was attentive as he told me not to tighten the connection of the new tank too tightly. It needed only to be snug, not
locked down. My mind was alert. The buzz and clamor of the lab
was beginning for that day. There was a life to it. Announcements
blared over the loudspeaker: "Positron filling is finished." "Injection is complete." "Tuning is complete, experimenters please
acknowledge." "Irving Johnson, line 8 please. Irving Johnson,
line 8." Metallic chimes tolled. Regular, repetitive. It seemed to
me like layers of activity. If you listened carefully to the exchanges

on the lab-wide intercom, you could keep up with what was going on in all the different sub-functions of the lab. That, I was to learn in time, was an important part of an operator's job. Coming to that bit of knowledge, however, was not straightforward. There were many members of the lab, and many opinions as to what was important and what wasn't, what was true and what wasn't, what was real and what wasn't. My first lesson in this regard came as the aforementioned operator and I returned to the control area. As my first tutor walked away, a second operator came up to me and asked what I had just done. When I told him who I had gone with and what we had done, he leaned over and told me not to listen to a word that guy said. I remember thinking that learning about the lab wasn't going to be easy.

I liked working on the experimental "floor," a cavernous room the size of a football field with a labyrinth of machines and equipment staking out space on it. I had a small desk in a room right off the "floor," but I spent almost no time at it. All the operators' desks were there, up against all four walls in a circle.

As the weeks went on, I relaxed my dress. Blue jeans and a T-shirt was my standard outfit, and I began to let my hair grow out, although I had a long way to go to match some of the other operators. That I didn't ride a motorcycle put me in the distinct minority among my new peers.

After a few months, I became responsible for several routine tasks. There were pieces of equipment that had to be kept cool with liquid nitrogen or they would suffer catastrophic failure. Several tanks of a variety of gases had to be replaced regularly. During my shifts, I would make rounds with a clipboard and record the various pressures and levels and note when I filled a liquid nitrogen dewar (a kind of flask) or changed a tank. The liquid nitrogen dispenser made a harsh screaming sound that was terrible to hear late at night. To have to go out in the freezing cold and get a new tank of oxygen was awful. I hated this part of the job, so ritualistic and mechanical. It felt like some kind of punishment.

My initiation continued as the operators, the operations manager, and the assistant operations manager told me about various aspects and systems of the lab. It was complicated. Not much was written down. I learned how to give a safety tour, and how and when to check and mark down important readouts. I was learning what the "beamline equipment" was, and how and when to talk to the storage ring's operator over the lab-wide communication system.

In my excitement at being in this new and strange world, I began to explore the various depictions of the relationship between the operators and the scientists at the lab that I was hearing about. The operators and the scientists, it seemed, inhabited different worlds. One operator told me that he liked to work the night shift because fewer scientists were around and he could just "set up and run the operation" without having to be embroiled in frustrating conversations about everything. Another operator who had been in the military told me that, much to his disappointment, he found the relationship similar to that between officers and enlisted men (Field Notes, Book 2, 2/19/93). I was given the password to an early form of a chat room in which operators wrote their impressions of the lab, their work, the scientists, and lab management. The scientists did not know of this log, or, I was told, of this capability of the lab's computing network. The remarks were, for the most part, derogatory and pointedly critical. I wondered why the operators felt this way about the scientists, and why they felt the need to hide their thoughts.

I asked some of the operators about how they had acquired their knowledge of the workings of the lab. The key, they told me, was their experience with equipment in general. One operator told me that as a boy he used to disassemble the various devices in his house, such as clocks and sewing machines, then see if he could put them back together. When I asked if he was encouraged to do this by his parents, he assured me with a laugh that it was quite the opposite. When another operator told me that he also took apart equipment as a child, I asked him who taught him how to understand equipment. "Well," he replied, "the equipment teaches you. It bothers you that you can't figure out why it works, so you take it apart and, over time, the more you know how to take it apart, the more you know how it goes back together. You break a few things and as time goes on you get better at it." He then said "I think it's the same reason someone has a desire to know about words and how they are put together." When I asked him if he was saying that he was using a kind of scientific method to gain knowledge about equipment, he agreed thoughtfully. He then added: "It really depends on whether you have a mind of looking at the process of getting to the end of what you want or whether you just want to get to the end and you're more concerned with the science output or whatever its going to be . . . there are people who just don't really concentrate on the fact that each individual component of any kind of system could be considered a weak link if its not done right. I don't know if that is some kind of character trait, I just see that in a lot of people." When I asked him if he had some particular people in mind, he replied "Well, if you ask me, a lot of the scientists around here don't have any scientific method about them at all." When I asked him whether the scientists had taught him about how the equipment at the lab works, he replied, curtly, "I taught them." (Field Notes, Interview, 9/21/92)

As I worked on projects at the lab, I began to inhabit and perform the operators' "just do it" approach with regard to equipment. I felt uncomfortable when I asked too many questions.

There weren't many manuals for the equipment, and those that existed were kept in a fairly unorganized filing cabinet. I remembered what I had heard—that the equipment teaches you. I was beginning to understand how to act, what to do, and how to perform as an operator.

In my first real project, I was told that a particular set of circuit boxes had been malfunctioning, and that I should fix them. Given the basics of what the boxes were supposed to do, I was then left alone. Even though the designer of the circuits worked at the lab and was only a few doors away from me, I remember feeling that it wouldn't be right to go down there and talk to him. At times I was fairly frustrated because I didn't really know what to do. I knew, though, that I could invoke the model of understanding equipment that I had heard about—that the equipment teaches. It was just me and the equipment, after all. I simply tried things, even if I couldn't exactly explain why. I was prepared, if I were to break something, to say something like "Well, I was investigating the circuit, and I'm learning, and, hey, you have to break some eggs to make an omelet." That was better than asking questions. As it turned out, I "fried" a few chips, kept the fact that I had done so to myself, and eventually got the boxes working. I knew I would be able to justify my actions later to my boss and to the other operators. I knew that I would be seen as doing what operators do.

The "Scarlet O": Knowledge Invalidated

The operators' antagonism toward the scientists with regard to understanding equipment was clear. The operators did not take things that the scientists said with regard to laboratory equipment at face value. One operator summarized the general feeling when he told me that scientific training only teaches a person how to get locked into concocting an argument for whatever point one wants (Field Notes, Book 6, 12/6/93). In time, I came to better understand why the operators felt that their knowledge was being discounted, but at the same time appropriated, by the scientists. The lab was about to undergo a major upgrade, and the talk among the operators was about what was going to be done and who would be involved. I remember the tension in the air when one of the senior operators announced to the group that the upgrade was going to go the usual way—i.e., that the operators would be brought in at the end, not at the beginning, and would have to make everything work. The operators resented that they would not be granted access to participate in designing and discussing the upgrade, but that they would be called upon to make the equipment work in the desired ways.

Many operators understood the tag of "operator" as a taint, a stereotype imposed upon them by scientists who ignored their input into laboratory matters. Once you were cast in the role of operator, according to this line of reasoning, you were stripped of the credibility necessary to be a valid knowledge contributor in the laboratory. Evoking the unjust persecution of the "adulteress" Hester Prynne in Nathaniel Hawthorne's novel *The Scarlet Letter*, they spoke of "the scarlet O" they thought they must have on their chests, marking them as "Operators" and invalidating their knowledge (Field Notes, Book 3, 10/13/98). Given the view that it was the operators who actually had a scientific method of understanding equipment, the operators saw the exclusion of their voices in issues of laboratory development as a sign of hypocrisy on the part of those claiming to seek scientific knowledge of experimental equipment and an intrusion of politics into the lab. In line with this reasoning, the operators saw an asymmetry to the way mistakes were defined at the lab. When operators made mistakes, it was because they didn't understand, didn't pay attention, or didn't care, and the mistakes were never forgiven. When scientists made mistakes, it was because they were confronted with difficult situations, and those episodes were quickly forgotten (Field Notes, Book 5, 11/18/93). To the operators, the standard by which something was or wasn't considered a mistake was whether or not you were an operator. They were the underclass. Invalid. One operator who had published a paper in a scientific journal while working toward a Master's in mechanical engineering told me, in the context of a discussion about a paper I was working on, "I got my article published for my Master's. I just kept a low profile (about it here). Never talk to anyone about it because they don't want you to have a brain—just lug lead bricks." (Field Notes, Book 2, 2/19/93)

I wasn't used to this. Coming from an Ivy League school and the corporate world, I was accustomed to being unselfconsciously valid. Now here I was, an operator, carrying gas tanks and sweeping floors, with an opinion but no voice. I would cringe when a scientist would introduce me to someone as an operator. And the pay was so low. My starting salary at the lab was $23,500 per year. (I had negotiated it up from the offer of $22,500, arguing that I had almost attained a Master's.) Most operators made a bit more than that, the high twenties, while the scientists made $50,000–60,000 a year. At the time, there were about ten operators and eight scientists at the x-ray lab, and two machinists. Although most of the operators were single, a few had children to support. Many times I would check in at midnight for the beginning of my shift and wonder what I had gotten myself into. It's hard to describe if you haven't felt it. It's not always pushed on you, but the regular reminders serve to point out that it's always there in the background and can be brought out at any time. It feels heavy. It doesn't go away. I was an operator. A worker.

Reflections on a Way of Knowing

Why were the operators so disgruntled? Why weren't they listened to? What was going on? What should I do? What did this have to do with science? To hear the operators tell it, they were not being recognized as the valid knowledge producers they felt themselves to be. Were they right? Were they wrong? What could be done? I was still working full time at the lab and taking classes extramurally, and as I delved into the literature I hoped to find some answers. In my searching, something began to crystallize. I was coming across variations on a model of science that were well suited for including technicians as valid producers of knowledge. It seemed that, in bringing out the complexity of scientific and engineering practice, many historians and sociologists had found cases in which "technologists" hadn't simply applied already-discovered scientific principles. Rather, new knowledge about how to proceed had been derived from interactions with the equipment itself! In this vein, the historian of technology John Staudenmaier has argued for considering "technological praxis as form of knowledge rather than an application of knowledge" (1985: 120). In a study of cases in aeronautical engineering, Walter Vincenti has noted that full technical knowledge "can only come from individual experiences" (1990: 190). Indeed, many historians and philosophers of technology have seen science as not prior to technology (Cardwell 1976; Gamber 1995; Konig 1996; Lelas 1993; Molella and Rheingold 1991). Coming at it from the scientific "side," Evelyn Fox Keller has argued that the Nobel laureate Barbara McClintock did better science because she was patient enough and willing to learn from the material of her study and to take seriously "every component" in its own right (1983: 200). To Keller, McClintock was innovative precisely because this valuable component of the scientific method was, as the operators at this lab have claimed, lost on most scientists. The labor historians Stephen Barley and Beth Bechky (1993: 11) picked up on this notion in their study of laboratory technicians when they noted that "the lab staff took pride in their ability to see intelligible codes where novices (even scientists) saw no information at all." The operators at my lab were articulating a kind of knowledge production capability in line with what Edwin Layton referred to as "technologists doing science," with what Barley and Bechky described as how a "scientist-technician" works, with descriptions of the "engineering sciences" by Vincenti, Staudenmaier, and others, and even with Keller's description of McClintock's working method! In all these cases, analysts with doctorates and academic appointments support a model of technical knowledge in which experience with equipment or the material at hand is integral and necessary to the doing of science.

That was it! The operators were right, and scholarship about science was on their side! The lab just needed to understand that the operators were working in a valid scientific way; then there would be less animosity and better mutual understanding. In fact, Barley and Bechky pointed out that not recognizing that science has this aspect to it and that technicians are well placed to contribute in this regard is a prevalent form of mismanagement among science laboratories (1993: 28). Whalley and Barley (1997: 22–53) even said that technicians should leverage their importance and expertise by unionizing and demanding better compensation and more autonomy in their working lives. But something wasn't right. As I went back and forth between the laboratory and the STS department, as I carried out my tasks at the lab, and as I lay in bed at night, I sometimes wondered whether I had some secret information, some special knowledge that the scientists and the operators didn't have. What should I do? Who should I tell? Something nagged at me, though. It seemed too convenient. Too simple. The operators were right, and scholarship on science and technology was on their side? What if I hadn't come into the lab as an operator? Why did the scientists not see it this way? Around this time, my trajectory at the lab underwent an important shift: I was promoted to assistant operations manager of the x-ray lab. After my promotion, I came to more intimately engage these scientists against whom the operators seemed to measure their identity, capabilities, and worth. As I did so, my understanding of knowledge claims at the laboratory became more complicated.

Lab Hands: Introduced by Scientists

It had come time to tear down and rebuild half of the lab's experimental stations, and I was approached by the operations manager and asked if I would be willing to take charge of the other half of the stations to make sure that those operations remained intact while many of the operators concentrated on the rebuilding. I felt pretty intimidated—nearly all the other operators had been there longer than I had, and I was still in the midst of trying to figure everything out—but I took the assignment. Working with the experiments and the experimenters seemed to be my forte at the lab, and I liked the idea that I would be close to the research front. Getting an experiment working, getting it humming along—I liked that. I was good at it. After about a year in this mode, I was made an assistant operations manager. I began spending more and more time interacting with the scientific staff. As an assistant operations manager, I attended the weekly senior staff meeting (to which no operators were invited), was involved in hiring decisions, and attended meetings with other laboratories as a representative of our lab. I also spent time discussing the operations, problems, and the future of the lab with scientists in their offices and on the experimental floor. As I moved into these different forums, I came to an understanding of the scientists' view of who operators were and what they could know and do.

The scientists' offices were one level upstairs from the experimental floor. The hallway resembled any bland academic hallway. In contrast to the operators, the scientists usually wore collared shirts and slacks or khakis. None rode a motorcycle or had long hair. I did not pick up any animosity toward the operators. Instead, I got the impression that the scientists, in general, considered the role of operator important and valued it. They often said that no project really gets anywhere without operators working on it. They even referred to the operators as the "life blood" of the laboratory. The laboratory's director was very proud that the job of operator as he conceived of it was envied and copied at other laboratories. Who, I wondered, were these operators that the scientists so valued? Were they the same ones I knew?

It was then that the scientists introduced me to an entity I had not previously heard of: "lab hands." An operator with good lab hands, it seemed, could work with equipment in an intuitive and comfortable way. An operator with good lab hands had a feel for equipment that enabled him or her to sense problems and get equipment working. In my discussions with the scientists, it became clear that this was the essential quality that an operator must possess. Scientists would ask me which operators had good lab hands and which did not. There was not much talk about training operators to have lab hands, and everyone agreed that it was impossible to discern them through conversation. Instead there were slightly anxious periods during which the scientists waited to see whether a newly hired operator had good hands. It was a great relief to find that one did. In a magazine article about the lab, one scientist boasted: "Our operators really know their jobs. . . . They have an intuition for the machine that I haven't seen in any other facility. Some of the operators can tell if the magnets are working properly just by laying their hands on them to check the temperature." (Saulnier 1996: 14) Job postings for the operator position from this period specified that "a two year commitment to the position is requested" and that "experience is not needed but mechanical and lab skills along with a BS or equivalent in technical/scientific fields such as physics or engineering are desired." Here experience was separated out from an applicant's ability to do the job. What was needed was a certain kind of skill, an innate feel for how to proceed. One measure of whether an applicant possessed this kind of feel for equipment was to ask, in the course of the interview, how he went about repairing and maintaining his car.

This view of the operators seemed to be based on quite a different model of understanding equipment than the one that I had been introduced to by the operators. According to this new view, ability was apparently embodied, inherent, located in the hands rather than the head of the operator, and brought to bear in interactions with equipment, not developed through such interactions. When it came to lab hands, it seemed, either you had them or you didn't. The naturalizing aspect of these characterizations struck me. Operators were simply people of a certain type. They were intuitive creatures. They felt things in a holistic kind of way that couldn't be formulated or explained. They inhabited an inarticulable world in which the "laying on of hands" solved problems. Why were the scientists promoting these characterizations? I remember thinking how odd it was that these tough-talking, motorcycle-riding men were having "soft" and holistic attributes assigned to them. It seemed to me that they were having traits projected upon them that, in my experience, had been projected onto women and native peoples. They were "primitive" in the sense that they were "in touch with" their natural world—the lab.

As I came to understand the scientists' version of who the operators were and what they could do, I looked back on my time at the lab and recalled how I had been placed into such a role or identity. One time, when I was working with some of the other operators on a piece of vacuum equipment, a scientist came over and handed me an assembly that consisted of a silicon crystal clamped to a base by a spring-loaded mechanism. After telling me that the crystal was not diffracting x-rays properly, the scientist asked me to feel whether it was clamped too tightly. He knew I had worked with these assemblies before. Why wouldn't he loosen it and see if it performed better? Why had he come to me? In one way, I was flattered to play a part in diagnosing a scientist's problem. In another way, I felt taken advantage of somehow. I put my finger on the crystal and tried to slide it in the clamp. When these assemblies had worked for me, the crystals could slide back and forth in their clamp. This one didn't slide. "It's way too tight," I told him, and he nodded and walked away. Apparently this exchange bolstered the scientist's confidence in deciding what action to take before going through the roughly half-hour of time required to replace the assembly into the device from which it came. The scientist did not replace the crystal, but rather loosened the clamps. When I saw him an hour later, he gave me a "thumbs up" and told me that it had worked. Another time, I had used sandpaper to "rough up" (after spitting on it) the surface of a block of crystal silicon that was being used in a monochromator, a device that separates out a narrow band of wavelengths from the "white" x-ray beam that emanates from the storage ring. With a rough surface, this crystal performed differently and in certain ways better than a "smooth" crystal. From that point on, whenever a crystal had to be roughened, it was brought to me. When they brought the crystals, the scientists would talk about my calibrated fingers and my saliva as if they were necessary for the job. (I was told that in any publication my saliva would be referred to as "distilled water.") (Field Notes, Book 2, 4/23/97) In both of these cases, my "feel" for the equipment had been put to use. Another time, a scientist came up to me when a vacuum pump didn't seem to be working and asked me if any of the operators had "done the laying on of the hands thing" to assess whether there was in fact a problem (Field Notes, Book 2, 11/20/96). After I informed him that I hadn't but perhaps someone else had, he sought out other operators. Another time, a member of the scientific staff introduced me to a summer intern as follows: "This is Park. He's interested in monochromators too, but he's more hands on, interested in getting something working. I more just sit in my office and design things." (Field Notes, Book 4, 5/31/94)

What had happened when I had touched the crystal in its clamp? What did it mean for operators to be "hands on" or to "do the laying on of the hands thing"? In the case of my interaction with the scientist, was I using knowledge based on experience, or was I using an innate feel? I didn't know. I felt flattered that the scientist would stake future actions on a declaration from me, but I also felt the sting. Did he know how many times I had been in his situation? Did he know how I had learned what I did? I knew that what I did was the fastest and easiest way to deal with an interruption in my busy day. This was the dissonance. When the operators heard their services praised, they felt they were being praised for being something that they weren't and for things that they weren't doing. And what they really were doing was ignored or, worse, co-opted. To the operators, these compliments from the scientists were backhanded slights to their real abilities.

Each group interpreted the operators' abilities on a different plane. Where the scientists saw interactions between the operators and equipment as tests of innate ability channeled through the direct connection of their lab hands to the equipment, the operators saw such interactions as indications of the knowledge they had gained through previous interactions with equipment. Where the scientists saw operators working with other kinds of equipment besides laboratory equipment as markers of general skill, the operators saw such interactions as markers of an ability to learn. When I asked a scientist about the difference between a scientist and an operator, he said that of course any good experimentalist needed the same kind of skills as an operator, but a scientist also needed the creative ability to sort through scientific ideas and produce experiments that would be interesting to the field (Field Notes, Interview, 4/14/94). A scientist could do what an operator did, but not vice versa. Scientists, according to this view, had heads and hands; operators only had hands. The operators, on the other hand, felt that a major problem with the lab was precisely the fact that their heads were procedurally ignored.

The operators, too, were conducting identity work with regard to the scientists. The way the operators saw it, their heads were ignored by scientists, some of whom, at least, were the type of people who were unwilling or unable to learn from equipment in the proper way. Scientists were "educated fools" (as the machinist Walt Protas referred to them) who didn't understand the real world of scientific instrumentation. Their adherence to the textbook and their narcissism with respect to their own knowledge was what kept them from being legitimate knowledge producers.

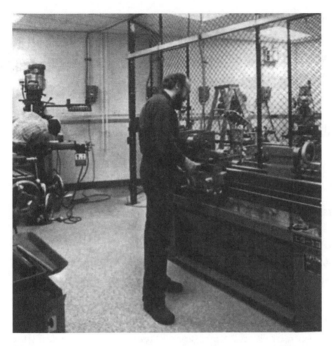

Walt Protas in the x-ray lab machine shop. (Batterman 1986)

The operators didn't know why the scientists were like this; they just knew that "some people were that way" (Field Notes, Interview, 9/21/92). As time went on, I wondered why this identity work was so prevalent at the lab. What were these characterizations, with their implications for knowledge production, really used for?

Epistemic Politics and Laboratory Labor

With regard to my study, I wondered what I should I do with these differing interpretations of laboratory practice and identity. I pondered over who was right, the operators or the scientists.

How could I understand what was going on with these competing identities and their implications for the nature of knowledge production? Indeed, what was the real story? Again I looked to the literature on science and technology. As I did so, I came upon writers for whom promotion of different models of knowledge production and performances of various technical identities of knowledge producers by actors in their studies were not taken at face value, but instead seen as moves made in an arena of authority and control. This resonated with me. Neither the scientists nor the operators were "right." Rather, they were each promoting and performing their own point of view, their own idea of who can produce technical knowledge and how it is produced. Mayr (1976), Barnes (1982), Layton (1976), Kline (1996), Anderson (1992), Faulkner (1994), and Cowan (1996) have all taken this methodological tack: the view that there is no one model of technical knowledge production per se, but that model production itself is the proper topic of study. That's also the lesson of Shapin's (1989) work on Robert Boyle's "technicians": it's not that the technicians weren't properly credited; it's that considering the hierarchical relations of scientific practice calls for a mode of analysis where the nature of a "scientist" or a "technician" is not taken as self-evident and prior to political relations.

What more can be said about the models of technical knowledge production and producers that are at play in the laboratory? What are they good for in this setting? How are they used? What work do they perform? How is identity work involved? The scientists' and the operators' depictions of one another and of their work helped them justify actions when dealing with equipment and also bolstered their claims for control of decision making and authority over the labor of the lab. The scientists' conception of operators and their abilities is based on a "feel" for the equipment that instrumentalizes the operators in several ways. First, the voice of the operators is treated as another scientific instrument, as in the case of the purported measuring of the temperature with hands. The scientists can listen to the operators as they would read instruments. Second, the skilled "lab hands" of the operators can be set to work in place of the scientists' own hands, leaving the scientists to operate in the creative realm of scientific ideas, technological innovation, and laboratory management. This kind of model implies that the scientists should be the ones who should control and direct the development of the lab; after all, they are the holders of technical knowledge. At the same time, this model helps the scientists justify actions in real time when working with laboratory equipment. The scientist who asked me to feel the crystal assembly could afterward both justify why he didn't replace the crystal in the assembly and also why he was rightfully my boss.

In the operator's version of laboratory practice, knowledge of the laboratory equipment was derived through the experience of working with equipment. They paid attention to detail, and they learned as they worked with different machines. This knowledge came directly from the interactions with the equipment; it did not depend on the accredited institutionalized knowledge of the scientists. In fact, the operators see themselves as deriving knowledge in a way that the scientists don't even understand. Their model usurps knowledge-producing capabilities from the scientists who do not spend as much time around the equipment and thus implies that they, the operators, should be rightfully more in control of developments at the lab.

Operators and Scientists: A Later Stage

As I carried my new awareness with me into my work at the laboratory, I found myself pushing "technical" arguments with a newfound zeal, for I knew what important work those arguments were doing. I learned when to push what kind of model of knowledge production. Meanwhile, my reputation as a serious lab member grew. As the life of the lab went on, however, I couldn't help but notice that something was changing. The attitudes of the operators and the scientists seemed different than before. The animosity and contention apparent in my first few years at the lab had diminished. There was now less argument over what an operator or a scientist could know. There was a certain calmness, lifelessness perhaps, in the air. It wasn't just me. I began to ask people if they noticed a difference. That the lab had changed became an accepted part of lab discussions. The culture of the lab, everyone agreed, was more stable now. I began to wonder why this was so.

From the beginning to the end of my working life at the lab, the number of experimental stations roughly doubled, and there was a trend toward designing stations to specialize in one particular type of experiment, especially protein crystallography, rather than supporting a wide range of capabilities. During this period, the scientific staff remained basically intact but the operator staff incurred significant turnover. And although there seem to have been some changes in institutionalized conceptions of the operator position that reflected the operators' views from the early phase, the predominant tone in the later phase was an acceptance of the division of labor implied in the operator-as-instrument model espoused by the scientists during the early part of my time at the lab. An indication that some of the early operators came to have an institutional validity can be seen in a job posting from 1996: "Job Opportunity: Research Support Specialist I. [X-ray lab] Operator. Requirements: Good computer, mechanical, and some electronic skills required as well as experience with maintaining scientific equipment. Good communication and people skills absolutely necessary. Bachelor's in physics or engineering or 2–3 years equivalent experience required." The most striking features of this listing are that the word 'operator' appears only in a secondary way and that the word 'technician' does not appear at all. The substitution of 'Research Support Specialist' for 'Technician' in the job title was seen by the early operators as a mark of the recognition and status they felt they had not received previously. The taint of the word 'operator' was being removed from the description of their work. Also, experience is considered differently in this posting than in the earlier years. It seems that now two to three years of equivalent experience is seen as in some way equal to a bachelor's degree. Here experience is seen as equivalent to institutionally certified knowledge in a way that it was not in the earlier listings. This too is in line with the operators' earlier conception of their work.

With these views of the operators from the early stage incorporated into management's view of the operator position, one might expect that operators in the later stage would more freely see themselves as doing properly scientific work and that their contribution would be openly valued by the scientists. But this simply wasn't true. In my last few years at the lab, despite the official name change and the altered job description, the operators were in fact seen as more removed than ever from the main business of the lab. The talk was no longer about the valuable "life blood of the lab," but rather about what was to be done about the operators' lack of knowledge and decision-making capability. In a switch from the first stage, the operators themselves shared the view that they were not sufficiently prepared. They themselves were frustrated by their lack of knowledge. Both groups saw the solution to this problem as a matter of the scientists training the operators. The operators were upset because they felt that the scientists hadn't taken enough time to train them properly to do their jobs. At this point, rather than claim that they needed time alone to work on and learn from the equipment, the operators put the onus on the scientists to give them the knowledge they needed to function at the lab.

The traditional division between scientist and technician was more entrenched than before. The operators did not consider themselves scientists. When I asked one why they did not, he replied: "Well, it's obvious, I'm not doing research of my own. I'm implementing designs and programs for other peoples' research." (Field Notes, Interview, 5/16/96) He then told me that the operators were not against the scientists, but that, the way he saw it, the operators and the scientists shared frustration with what they saw as an increasingly bureaucratic work environment. At this time, operator training was seen as one of the lab's most important priorities, and conversations about what an operator and a scientist could know were far less frequent and were increasingly seen as irrelevant. It was accepted that knowledge about the equipment derived from the scientists and should be passed on to the operators. Operators did not claim knowledge hard won from experience, and the scientists did not boast about the direct connection of the operators "lab hands" to the equipment of the laboratory.

Why did the conversations and the relations change? Why were the operators able to make such claims with such convictions during my early years? Why were the operators' and the scientists' knowledge-producing capabilities seen as more straightforwardly understood in my later years at the lab? Why did the emphasis shift from experience to training? The lab had grown larger and had become more bureaucratic, more institutionally entrenched. The equipment had become more stable, more simply present. One aspect of this growth seems to be that the epistemic resource of experience had a different cachet, for the operators and the scientists, from the beginning to the end of this period.

Epistemic Politics: Conceptions for Use?

In the time of early development at the lab, a good portion of the operators had taken part in building the equipment and could use appeals to this experience to bolster their claims of knowledge and control. As we have seen, experience is the cornerstone for the operators' claim to be knowledge producers. The fact that they were in on the ground floor of developing the equipment at the lab makes the appeal to experience in this setting a strong one. Some of the scientists had arrived at the lab later, around the time I had, after a significant amount of the lab's equipment was built up. Thus, they could not counter the operators' appeals to experience with credible claims of experience of their own.

As the lab developed, this situation changed. As new operators came in, they were in a much different position than the original operators who were leaving. The new arrivals came to a lab where most of the equipment already existed. They could not make the same credible claims to experience that the original operators could. Fixing a clock was one thing, but to have been working with the lab's equipment from the early days was a sturdy resource to call upon. The new operators were without this important resource and had no recourse but to defer to the scientists, who had not only greater accredited institutional knowledge but also now a history with the equipment at hand. The operators had little choice but to revert to performances of the predominantly accepted "lab hands" model and to publicly point out that their knowledge was dependent on the scientists. The scientists could now lay claim to the old operators' epistemic territory of experience with equipment better than the new operators could, because the scientists had now worked with the laboratory equipment for a longer time than the newly hired operators. The operators' and the scientists' claims about their knowledge-producing capabilities are not simply indicators of their true natures but rather performances that link the technical and the organizational and that privilege each group's status and authority over those of the other group. The battle for epistemic control was also a battle for workplace control between these two groups, and the terms and dynamics of this epistemic politics had changed.

An understanding of epistemic politics is the foundation for the project of this book. In order to further understand this project, it is worthwhile to explore two prominent concepts in the study of laboratory practice: Karin Knorr Cetina's "epistemic cultures" and Peter Galison's "trading zones."

In her 1999 study of particle physicists and molecular biologists, Knorr Cetina describes two different "epistemic cultures" that define each of the fields and determine what they, as technical knowledge producers, do. For Knorr Cetina, the large-scale, centralized, instrument-intensive field of particle physics employs a "classic" self-referential semiotic network of signs of objects based on "representational technologies" (1999: 80). Through three "ethnomethods" particular to particle physicists, "unfolding, framing, and convoluting," the physicists "articulate" their "internally referential system" to produce the natural world. What does this mean? For Knorr Cetina, particle physicists inhabit a world of complicated instruments in which simulations and predictions are integral to making sense of how that equipment registers the "real" world. Knorr Cetina sees the physicists as comfortable with a complicated repertoire of comparing and mixing "simulations" and "experiments." Instrument simulations based on calculations are not just tests. Instead, "convoluted" tangles of calculation, instrument simulation, and instrument "readings" are "expedient in creating experimental outcomes . . . in a world that refers back upon itself and seeks recourse in manipulating its own components" (ibid.: 77–80). In this respect, Knorr Cetina sees a world "marked by the loss of the empirical" (ibid.: 79). She contrasts this to the "small science" of molecular biology, which appears to her to "rely on maximizing contact with the empirical world" (ibid.: 79). In molecular biology, moreover, this contact with the empirical involves the "sensory" and "acting" body of the scientist. Small-scale manipulation at the laboratory bench is the hallmark of this field and is also grist for the epistemic mill of producing accounts of the natural world. Knorr Cetina describes the kind of researcher such work requires as follows:

[In] the molecular biology laboratories studied . . . a scientist's sensory skills, in the holistic sense, were continuously required. They were implied when some participants were said to have a "golden touch" or to be "excellent experimentalists." When students were recruited to the laboratory, older members watched for these qualities and recruited students accordingly. Conflicts arose when someone, highly recommended by an outside scientist, turned out to be 'hopeless in the lab" and "incapable of getting an experiment to run." (ibid.: 96)

For Knorr Cetina, the molecular biologist's body is an "information-processing tool." "Many scientists," she notes, "feel it is impossible to try to reason through the problem or to pick up the important clues from oral or written descriptions. In order to know what to think, one has to place oneself in the situation. The body is trusted to pick up and process what the mind cannot anticipate." (ibid.: 96) This information-processing tool is a primary component in the mechanism of blind variation at the laboratory bench, the fundamental mechanism by which molecular biology produces knowledge. Knorr Cetina encapsulates the difference between particle physics and biology as follows: "[In] HEP experiments, it is not the phenomena themselves which are at issue, but rather their reflection in the light of the internal megamachinery that envelops and tracks down physical occurrences. In the molecular biology laboratory, in contrast, the phenomena assert themselves as independent beings and inscribe themselves in scientists' feelings and experiences." (ibid.: 79)

According to my analysis, it is a mistake to take such assertions at face value. In the case of the operators in my lab, arguments for such a knowledge-producing capability were an effort to distinguish and privilege the operator's mode of knowing over that of the scientists who controlled the working life of the lab. Conversely, the scientists portrayed the operators in such a way that the scientists themselves were naturally the source of, and in control of, any "knowledge" the operators might produce. Casting the operators as skilled and intuitive, with a feel for laboratory equipment, erased them as knowledge producers and justified the control of their labor. The operators fought against this with identity work of their own, portraying the scientists as booksmart "educated fools" incapable of properly learning from experience with regard to laboratory equipment, thus placing them properly in control of laboratory development and work. At the lab, these depictions were used to privilege each group over the other as properly in control of laboratory practice.

Trading Zones and Epistemic Politics: Accounting for Antagonism?

An accounting of the interactions between subcultures in physics underpins Peter Galison's (1997) approach to understanding the scientific enterprise, and understandings of what technical practitioners are and what they do underpin that accounting. It is not a thoroughly essentialist reading, though, as Knorr Cetina's was. For Galison, these subcultures are "finite traditions with their own dynamics that are linked not by homogenization, but by local coordination" (ibid.: 803). This local coordination takes place in "trading zones." Galison describes a trading zone between two groups as follows: "What is crucial is that in the local context of the trading zone, despite the differences in classification, significance, and standards of demonstration, the two groups can collaborate . . . even when the significance of the objects traded—and of the trade itself—may be utterly different for the two sides." (ibid.: 803) Closest to the focus of the previous chapters of this book is Galison's account of relations among engineers, experimentalists, and theorists at the MIT "Rad Lab" during World War II. In the Rad Lab, the groups were "under the gun" to get on with the job of building radar. There was a particular space, room 4-133, where physicists and engineers would "trade." The primary example given is how the interaction between the theorist Julian Schwinger (who also first proposed the concept of synchrotron radiation) and the "'good enough" and input-output engineering culture of the Rad Lab resulted in more "practical" equivalent circuit representations of microwave theory that were brought into the practice of designing radar transmitters and which affected future abstract theorizing (ibid.: 821).

In the picture painted here, neither the traditions and practices of theoretical physics nor the traditions and practices of engineering are seen by Galison or by the practitioners as having epistemological primacy, and who is or should be in control of the project of developing radar is not an issue. In the case of the operators and the scientists at the x-ray lab, the operators used what resources they could (primarily their experiences with the equipment of the lab) to distinguish and privilege their way of knowing over that of the scientists, and the scientists resisted. The experimental floor was not a zone for mutually beneficial trading; it was a forum for antagonistic epistemic-political performances. Crucial to the resistances was how each group staked a claim to and was ceded (or not) epistemic territory that could not (or would not) be accessed by the group toward which the resistance was directed.

In exploring how claims and counterclaims of knowledge and knowledge-producing ability were handled between the scientists and operators at the x-ray lab, I came to see how important it was to not presume what kinds of practitioners the laboratory members really were or how they really went about knowing what they claimed to know, for these were the achievements that were precisely at stake in the dynamics of laboratory practice. This point is crucial in exploring contingency in science, and it guided my work through the rest of my study as I became involved in further episodes of decision making about the operation of the storage ring, the development of experimental instrumentation, and the nature of experimentation at the lab. Indeed, considering interactions between different scientific subcultures as taking place in forums of epistemological and political agonizing is the means by which this book explores how laboratory practice changed and the implication of that change for the products of science.

4   From Ion Trapping to Intensive Tuning: The Particle Group, the X-Ray Lab, and a Re-Negotiation of "Normal" Running

I was settling into my night shift as the x-ray operator on duty, checking the various temperatures and pressures on and in different pieces of equipment at the x-ray lab area. I would be the point man tonight, helping the various experimenters and making sure the equipment was working. Things seemed to be running smoothly. The low buzz and hum of the lab can be relaxing. As I was noting the temperature of a thermocouple on the F-line, a voice crackled over the lab-wide intercom: "X-ray lab operator line 1 please. X-ray lab operator line 1."

I picked up the nearest receiver and identified myself as the x-ray lab operator. The storage ring operator on the other end, calling from the CESR control room, said "We're experiencing ion trapping" and told me the run would have to end. I had a vague idea as to what "ion trapping" meant. Some of the vacuum pumps around the ring used ions to attract particles from inside the ring. Sometimes these pumps somehow interacted in a bad way with the beam inside the ring, and the run was compromised. When this happened, there was nothing to do but "dump" the now destabilized particles, refill the ring, and start the run over. I wondered who storage ring operator meant by "we." I told him I would have to go around and check with the experimenters who

The CESR control room in 1998. Richard Eshelman is at the controls. (P. Doing)

were working at the x-ray lab's stations. I stalled for a couple of minutes, then called for the storage ring operator: "CESR operator line 1." When he picked up the phone, I told him that "we" were ready. Then, over the lab-wide speaker system, all present heard these announcements:

> "Experimenters, I would like to end the run. Please acknowledge when ready."
> "CHESS is ready."
> "Particle group is ready." [The particle physics group monitoring the output of their detector had also been told on a direct line that the run would have to end.]
> "Thank you CHESS. Thank you CLEO."

The synchrotron operator then "dumped" the electrons and positrons that were orbiting inside the ring, ending after only 10 minutes an experimental run of high-energy particle physics and

The x-ray lab's control area. (Rice and Fontes 1999)

synchrotron x-ray experimentation that was scheduled to last 60 minutes As a beginning x-ray lab operator, I presented this decision to x-ray lab experimenters as it was presented to me, as a drastic but necessary response to an unsolvable technical problem (Operator Log, 11/14/94).

On any night, a myriad of activities take place at the lab. On the experimental floor, a wide array of visiting researchers, mostly young professors and graduate students, work at the x-ray lab's stations. X-ray lab operators monitor the situation, helping, sometimes teaching, and in general "supporting" these experimenters. For the x-ray lab to be open and operating, an x-ray lab operator must be on duty. The x-ray lab operator is responsible for the safety aspects of the lab and for the initial response to difficulties with x-ray lab instrumentation. Scientists employed by the x-ray lab also conduct their own experimental work or collaborate with visiting scientists on experiments. Meanwhile, on the

The inside of an x-ray lab experimental "hutch," circa 1997. (P. Doing)

floor above, members of the particle group known as the CLEO collaboration watch a series of computer monitors that flash indications of the results of the collisions of electrons and positrons inside the particle group's detector. This raw streaming information is stored for later processing and analysis. Typically, particle group graduate students monitor the output at this stage. Next to that room, again a level above the experimental floor, the CESR operator injects electrons and positrons into the storage ring while simultaneously making a variety of adjustments such that as much "current" (amount of electrons and positrons) as possible could be injected into the ring. Once the electrons and positrons are injected into the ring, the operator continues to make adjustments so that these particles circulate in the machine in the most efficient manner possible and collide with each other at just the right place.

After injection, an "experimental run" officially begins. During experimental runs, all the researchers at the lab—members of the particle group and of the x-ray group alike—pursue their different experiments at the same time. While members of the particle group monitor the results of electron-positron collisions in their detector, x-ray experimenters work in the experimental stations on the laboratory floor, exploring the interactions between the x-rays emanating from the ring and a variety of materials ranging from semi-conductors to proteins and viruses. Over the lab-wide intercom, the machine operator contacts both the x-ray area and the particle group's control room to announce that the "run" has begun. These "HEP [high-energy physics] runs," as everyone at the lab called them, were generally scheduled for an hour. At the end of the hour, the machine again contacts the x-ray area and the particle group to announce the end of the run, "dumping" the remaining particles from the ring and injecting new ones for another run.

During laboratory running periods, this pattern of operation repeats continually for four or five days out of every seven. During these periods, the lab runs 24 hours a day, with the staff and the experimental groups "pulling shifts" round the clock. On the other two or three days, the particle group conducts "machine studies"—experimental tests that are used to gather new information about the capabilities of the machine, the detector, and the x-ray lab. "Machine studies" time is also used to test new components of the machine, the detector, and the x-ray lab. Between these two functions, HEP runs and machine studies, the laboratory is staffed, open, and operating 24 hours a day, seven days a week, 52 weeks a year, although this pattern is broken when the laboratory interrupts, sometimes for months at a time, to perform upgrades to the facility. On this night, many runs were being "dumped" early.

The dripping stalactites on the ceiling of the tunnel gave the impression that it was alive. The many-legged dust-colored insects that occasionally darted out from under pieces of equipment and the rumored tunnel rats filled out the ecosystem. A certain anxiety visited me every time I went in the tunnel. We were repeatedly told of the dangers of the tunnel and the consequences of a mistake. The worst thing was to be the last one out there, delaying the start up of the machine. Well, the worst thing was to break something, to bring the machine down. Walking among equipment I did not understand, I pretended that it was natural that I was there. It's true that our equipment was out there. Pipes attached to the machine in which the electrons were accelerated in a disjointed circle "collected" the x-rays that were discharged as each particle's trajectory was forcefully changed by magnetic fields. They might have been our pipes, but it was the particle group's tunnel. Like work gangs on some kind of futuristic highway of steel, copper, and aluminum, these groups of men would be surrounding and climbing on different sections of the machine as I passed by. I remember how I felt obligated to explain what we were doing. They were always skeptical. They called us the "country club." We were the dilettantes, the tourists.

The big news around the lab at this time was the competitions between the particle physics group at our lab and another group at another lab. A big new detector would be funded for B-meson studies with a new ring. It would set the lab up for the next ten years, maybe more. I heard that a lot of the guys working on the proposal had come from the Superconducting Supercollider project. We were a lot smaller than the other lab in the competition, but we had a chance. We could do it for a lot less money, and we knew we could do it just as well. Well, I used to say "we" when I talked about it to people, but it wasn't "we" really. We at the x-ray lab had very little to do with it. It was the particle group's world. There was a buzz, though, an excitement in the air all through the lab. We were going to get the B-factory.

Just Our Luck: Ion Trapping as Operator Error

As a new operator, I was in much the same position as the majority of the x-ray lab's users. I had been working at the lab for about three months and had received a torrent of information concerning the equipment and procedures. Comprehending the technical subtleties of the ring was somewhat of a stretch for me. As I struggled to understand these strange new surroundings, I had no recourse but to accept the interpretation of ion trapping as implied by the synchrotron group's operator. I would convey to the users of the x-ray lab a sense of disappointment and apology, but not a sense that things could be otherwise, that this reality could be questioned. With time I became more familiar with more aspects of the lab. As I did so, I came to question my understanding of the effect known as "ion trapping." I came to see others doing so too. One indication that there could be more to the story was the seeming correspondence between periods of heavy ion trapping and the shift schedule of the machine operators. When experimenters who had noticed the same thing would come up to me and ask me why the trouble seemed to start at 4 p.m., for example, and to continue until about midnight, I would explain that it was a very complicated machine and that some operators were just better at running it than others. At this point my interpretation of ion trapping had changed. It was still a technical problem, but it was not an intractable one. A talented machine operator could avoid it. As this version of ion trapping entered into my repertoire of explanation, the x-ray lab's users would occasionally curse their luck at being scheduled at the same time

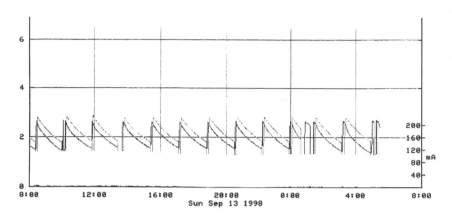

A printout of the synchrotron scoreboard. The *x* axis is the time in military time. The *y* axis is the amount of electrons (dashed line) and positrons (solid line) (in milliamperes) that are in the storage ring. Each run begins with injections that result in vertical lines. As the run progresses, the amount of electrons and positrons in the ring declines as particles are lost through collisions and other factors. If running were perfect, the scoreboard would show a "sawtooth" pattern all the way across. In this case, a fairly perfect sawtooth pattern of running is interrupted shortly after midnight—the time of a shift change in machine operators. Instead of more than four runs per 8-hour period, as had been the pattern, only two such runs were delivered in this shift. X-ray lab experimenters would point to such scoreboard output to bolster their point that the actual operator mattered to the running of the machine. The date of this record is 9/13/98. This log was stored (in real time) in a database that can be accessed through the laboratory's website.

as a less-than-stellar machine group operator. At the start of any running period, I would make it a point to find out which synchrotron operators were scheduled while I was shifted in order to get a preview of what the running was going to be like. Again, I presented the situation to the x-ray lab experimenters, and it was taken, as unfortunate but necessary. This reality also could not be otherwise.

Protein crystallography was becoming a bigger and bigger part of work at the x-ray lab, and the protein crystallographers were the most difficult of the users at the lab s about. They would come around a corner with a distressed look. "Our beam is gone!" they would inform me without elaborating. Their tone implied that it was my fault. It used to work me up a little. Didn't they understand how complicated the machine was? There were at least a dozen ways in which the x-ray beam might drift slightly out of alignment with their apparatus. They didn't know and didn't care. It was my problem, and they expected their beam. The physicists usually tried to empathize with the complications involved in running the storage ring. It was easier to explain to them that the machine was running poorly. I remember how the biologists seemed to take offense. Usually, a few minutes after a complaint from the biologists, the call would come down from the machine group operator: "Ion trapping!"

It was Christmas Eve day. The lab always ran 24 hours a day on holidays and new operators pulled these shifts. It could be very peaceful. The quiet hum of machinery, the low buzz of data collection. There were few calls from the outside. We were outcasts, underground in our own world. I had been at the lab six months and I was starting to feel comfortable. I was alone, in charge, and everything was working. Suddenly, a sharp, shrill beeping sound crackled through the air. In one of

the chambers in one of the pipes that were attached to the machine, the
pressure was rising. That probably was because of a leak. One of the
beryllium "windows" that allowed x-rays to pass through it but sep-
arated the vacuum in the machine from the atmosphere of the x-ray
experimental stations was probably corrupted. It was time to go into
the tunnel. But no one else from the x-ray lab was here. It was just me.
There was only the synchrotron operator, and he had to stay in the con-
trol room. There was no way I could do it alone. I called the x-ray lab
operations manager and after quizzing me, he came in.

Well, this was real tunnel work, just him and me. We told the
machine operator what was going on, the machine was shut down, and
we went in. We sealed off the appropriate sections and broke the vac-
uum. Sure enough, the window was cracked. We wrestled a new one
into place. We were like surgeons, with our white gloves and our seri-
ous, almost somber demeanor. It was in. Fire the pumps back up. It's
holding. The vacuum's coming down. Right near the end I was shocked
with fright when one of the cooling water fittings broke and an explo-
sion of water sprayed all over us and the equipment. But my mentor was
calm, it was outside the pipe. It would dry. Everything was OK. Nothing
was shorted. We finished our work. We returned our keys to the control
room, explaining that everything was OK. It's true, I was just the assis-
tant, but that trip into the tunnel changed me. From now on I would take
a different tone with the machine people: more confident, more asser-
tive. I would hold my head higher when among them. (Operator Log,
12/24/94)

Ion Trapping: The Robertson Effect

It was a rough night at the lab. There were many episodes of ion
trapping. Tensions were rising, and users of the x-ray lab were
getting exasperated. As the next operator came in to relieve me

of my shift and I told him about the short runs and frequent dumps, he said "Aaaah, the Robertson effect." When I asked what the Robertson effect was, he told me that there was a particular machine group operator who had the ability to drive the machine to higher currents. The x-ray operator telling me this had been at the lab for a while and had been a manager at different points. He spoke to me with a slightly condescending tone. I was informed that Robertson (a pseudonym) was their "top gun." My colleague explained to me that for the particle physics group luminosity (that is, the number of particle collisions per unit of time) was critically important for the survival of their experiment. He told me that Robertson was the best machine operator at pushing the machine to higher and higher currents. Ion trapping was simply the result of the machine's being pushed that hard. Robertson wasn't incompetent. He was the best at his job! When he was on shift, we at the x-ray lab just had to live with periods when runs were shortened as a result of ion trapping as Robertson tried to find ways to inject more particles into the ring. When he left his shift, the machine was very often running at higher currents (more particles in the machine) than when he had arrived. In the x-ray lab's logbook, where I had previously kept a detailed record of every run as it ended early and was dumped, my replacement simply wrote that the "Robertson effect" was controlling the machine and that he would note the time when good running resumed (Operator Log, 4/17/95).

Now the situation had become more complicated and more difficult for me. I didn't feel right relating this interpretation—that the machine was being purposely put into a mode where runs would be short and that "ion trapping" was a result of this activity—to users of the x-ray lab. For the most part, an increase in ring current was not critical to x-ray experiments. The fact that a synchrotron was producing the x-rays made for plenty of power. What x-ray users needed was a steady beam and time to collect their data. After such periods of short runs and many dumps, it was often necessary to realign the x-ray experiments because a new "sweet spot" had been found for the particle beams that allowed the injection of more particles. Therefore, to explain to users that rough running and realignment of their experiments was a matter of a lab policy of which they had not been informed, or something that our x-ray lab at least had no control over, seemed to me at the least awkward and at most disingenuous. This situation created a tension on the laboratory floor and a parallel tension in laboratory meetings. In time, "The Robertson Effect" became a prominent topic of conversation throughout the x-ray lab.

The competition was heating up. The word was that we were definitely in the hunt. The other group was proposing technologies that hadn't been used before. Our proposal relied on an extension of what we had already put to use at our lab. And we were coming in at half the price! It was a bit of a stretch, but we had shown that we could pull it off. And we had a history of pulling things off. The anticipation was palpable. We really were going to get the new B-factory!

Working Toward a New Order

"The Robertson Effect" was discussed in meetings of the x-ray lab's senior staff. Other than me, the participants in those meetings were the lab's operations manager, the lab's scientists, and the lab's director and assistant director. These were the meetings at which management decisions regarding the development of the lab were made. The topic of machine running was always discussed as the first item at a meeting. The following are summaries of assertions and considerations that I recorded at the x-ray lab senior staff meetings that I attended as the assistant x-ray lab operations manager from March 1995 to April 1996:

X-ray lab needs to keep better records. Every week the particle group forgets what happened last week. They think everything was OK. This place has no institutional memory. We need a down to the hour log. The operators have to keep meticulous notes. We'll present it in the machine group operations meeting, but we have to be clear and honest about our own problems. It has to be clear what was our fault and what was the machine. Don't be heavy handed about it, just put the numbers up there so they can see what's going on. Listen, they don't want to be down all the time either.

X-ray lab needs to be somehow experimentally disengaged from the machine and particle groups. What we need is a feedback system! That will take care of this whole thing. Its simple: set a range of allowable motion for the particle beams as they pass through the x-ray lab region and lock it in. If the particles stray out of range, have a feedback loop kick the magnets to steer them back. It could all be automated! Nobody would have to think about it anymore. But whose monitors should we use to detect where the particles are, theirs or ours? Well, ours monitored the x-rays, theirs monitored the particles themselves. Do ours work? Do theirs work? How do we know that they work? We can't tie the operation of the machine to flaky monitors. The particle group will never strap themselves to our x-ray lab monitors. Yeah they will, it will help them, it'll keep us out of their hair. The thing is they don't trust our monitors.

They say our monitors are flaky—are they? Who's in charge of our monitors? How can we have monitors that we don't trust? That's absurd. We need to be able to move them remotely to test them, we have to schedule time to go in the tunnel and make sure the motors are working and calibrated. We could move the monitors or move the beam. For the machine group to trust our monitors we need to calibrate our monitors to their beam monitors and then to the signals that power the magnets. We can't check them without beam time. We have to propose machine time for this—but our house has to be in order first!

From these senior staff meetings we at the x-ray lab developed a strategy. For our reports at the machine group operations meeting and for any future hardware feedback system, we needed to put in a major effort to prove to the particle group that our position monitors could be trusted, that when we said the beam moved that means it moved. To do this we knew we needed to include them in the calibration process. Our operations manager felt that he could work closely with the CESR operations manager, who he felt was at least not hostile to our cause. The operations manager was put in charge of the project from our end. His mission was to get our own house in order and work with their operations people in order that our monitors could be trusted.

One day, the x-ray lab's operations manager was out. It was a big day for me. I was to take his place at the meeting where synchrotron "machine time" devoted to future laboratory developments would be allocated. There I was, in a room full of world-class physicists, some of whom had built the machine 20 years earlier, standing up and giving a presentation about why we at the x-ray lab needed dedicated storage-ring time to calibrate our position monitors. We needed to run the machine and read our position monitors, and then gain access to the tunnel to check and calibrate the motors that controlled the physical position of the beam position monitors, in order to cross check our readings. My presentation was short, with not much discussion. I was told to talk with the head of the machine group after the meeting. I found him in the hallway, talking to other members of the machine group. After waiting to speak, I began to explain our need. I had never met the machine group leader. Before I was through, the group leader looked up, over my head, and, with a somewhat annoyed tone, asked another machine group member

"Is this really important?" I was allowed to finish my pitch, but we did not get any time that week.

Meanwhile, the situation with the biologists was worsening. At any given time, there were two or three protein crystallography groups running a total of six to nine experiments on the x-ray lab's beamlines. Moreover, the time allotted to these groups was getting shorter and shorter, which meant that any difficulties were magnified. Some groups only had 12 hours to work, and 24 hours was not uncommon. The apparatus for these biology experiments was very similar even on different experimental stations, and the performance of this equipment was monitored and supported by an organizationally separate group within the x-ray lab. This bio-x group (called MacCHESS, the 'Mac' standing for 'macromolecular crystallography') fielded feedback from groups who conducted these experiments at the lab and also pursued their own biological research projects. I had begun to take on more responsibility at the lab and was directly involved in overseeing the performance of the x-ray optics associated with these experiments. Relations between the bio-x group and me were worsening. Week after week in the general x-ray lab operations meeting the bio-x group would explain how much time in their experiments was lost, during the previous week, because of x-ray beam motion. It was frustrating for me because it wasn't clear how much responsibility an experimental group should rightfully bear for re-aligning their experiments in the course of a running period. It seemed reasonable to me that they should make periodic adjustments as their experiment went on, but they didn't see it that way. At other synchrotron sources they could go 12 hours or more without re-aligning. Why should they have to do it here? This situation was further confounded by the fact that that the beam motion could be coming from two fundamentally different sources. It was possible that the particle beam was moving in the ring, but it was also possible that the x-ray optics that monochromatized the "white" x-ray light coming from the machine were distorting or otherwise flexing as a result of the heat load from the powerful incident beam and thus causing the x-rays that made it to the experiment to move around. Since we couldn't trust our beam-position monitors located "upstream" of the monochromators, we couldn't clearly distinguish these two cases. Any episode of beam motion, then, would require time-consuming and often controversial investigation, with the biologists indignant that an investigation had to be conducted at all.

There was an deflated anxiety in the hallways of the lab. The particle physics group had not been awarded the big project! The award had gone to the other lab! There was a sense of dismay. Was it anger too? A classic case of politics. It was a presidential election year. The region that was granted the money was a known high-technology area with many important electoral votes. It was an obvious pork-barrel job. Those electoral votes were going to come in handy, and they were worth the extra money. How could we compete with that? Who was it going to help to pump money into our region? Technically both laboratories could do it, and our price tag was half of theirs. If they were chosen, it was because of politics. Well, we knew we lived in a political world, and everyone knew that it wasn't the first time politics had intruded into scientific research. Several physicists at the lab had been through the crushing cancellation of the Superconducting Supercollider only a few years before. The mood was down, frustrated. These were good technical ideas—nothing was wrong with them, and they would lead to good science. Things were going the wrong way, for the wrong reasons, and it was out of our control. The sense of loss was palpable.

Toward a New Understanding

We understood the kind of pressure that the particle group was under. Theirs was an experiment that compiled statistical data over months and years. For them, the more collisions between particles that they could record, the better their experimental results would be. Accordingly, they were always looking for ways to get more particles into the ring. If, during the operation of the machine, clues were deciphered that pointed to a way to increase the number of particles in the machine, it was considered a good risk to try something on the spot. Sure, there was risk, but it was worth it. We didn't want to be seen as self-serving complainers who didn't understand the big picture, but we did want to make our needs known more strongly.

We always gave a report at the end of the lab-wide operations meetings. Whereas before our reports were cursory, we now put up our new "accounting chart" that simply stated the amount of scheduled beam time that we used for experimenting and the amount of scheduled beam time that we did not, with a breakdown of the unused time into categories of our own operational problems and operational problems of the ring in general. We didn't spend much time elaborating on our own problems or pressing for answers with regard to ring operation. We just put the figures up for all to see. As time went on, this chart became an accepted and standard part of the lab-wide operations meeting.

Meanwhile, all of us, the particle group included, dreamed of the feedback system that would solve this problem once and for all. We imagined a time when we wouldn't have to argue or even think about these matters; they would be taken care of in hardware. The particle people could do what they wanted, and

Week 31
July 31 - eve. Aug. 6, 1997

Scheduled HEP shifts   18.5

Delivered HEP shifts   13.5

Comments:
1. 7/31 HEP started late lost 1/2 day shift and 1/2 eve
2. 8/1 Dipole magnet problems lost all day shift and 1/2 eve shift to realign
3. 8/3 Magnet problems lose 1/2 day shift
4. 8/5 Access and Filling problems lose 1/2 day shift and 1/2 eve shift.
5. 8/6 Machine studies taken lose all day shift

User Shifts Received

Station	Proposal	Shifts rec'd	Shifts sched.	Comments
A1	P712	1.5	3.5	
	EM203	5	6	
	P746	3	3	
	P669	4	6	
A2	FS149	13.5	18.5	
B1	dev	1.5	3.5	
	P671	12	15	
B2	ops	13.5	18.5	
C1	dev	13.5	18.5	
D1	FS181	9.5	12.5	
	FS131	4	6	
F1	P615	12	15.5	Computer prblms
	P693	1	3	
F2	P746	8.5	15.5	Piezzo problems
	EM221	1.5	3	
F3	P727	13.5	18.5	

efficiency 97.9%

A chart presented by the x-ray lab's operations manager at a meeting of the CESR operations group. "HEP shift" means High Energy Physics shift (the time in which both the x-ray lab and the particle group conduct their experiments). Each shift is 8 hours. In this report, 18.5 shifts were scheduled, but only 13.5 of those shifts were usable by the groups. Out of just over 6 days of running that were scheduled, about 4.5 days' running was "delivered." The efficiency percentage at the bottom refers to problems that x-ray lab had while there was usable HEP running. This chart was presented as a simple record of the previous week's activity. (P. Doing) it wouldn't affect us! The particle group would no longer have to put up with us on their back all the time. If only there could be such a system, we could stop worrying about this. I remember another machine group operations meeting: I was scanning around the room—the head of the particle group, the head of the x-ray lab—my eyes came to rest on a former head of the particle group seated in front of me. He was a man who carried great technical authority at the lab. He usually ended the conversations. In the meeting I could see him paying close attention to our x-ray lab operations manager's presentation. The topic was, again, loss of x-ray lab experimental time due to particle beam motion. Our operations manager was putting up the viewgraph of our accounting chart. The former particle group detector's yellow pad was in view directly in front of me, and I couldn't help but notice when he wrote "get somebody serious on the x-ray lab feedback."

After a few months, a solution was proposed, although it was not the hardware system that was dreamed of. For the x-ray lab, the most difficult part of the current situation was the uncertainty. To not know when experiments were going to have to be re-aligned and for x-ray lab users to never know when the beam was going to go away caused a constant level of anxiety. Maybe if we could at least be certain of some period of smooth running, then we would be willing to accept other periods of rough running. From this concession, a new policy of machine operation was born. Soon a new kind of schedule was released that described new running modes that would be in effect for any running period. It was agreed that there would be periods of "intensive tuning" during which the machine operator was under no constraints in terms of beam position or run length, periods of "intermediate tuning" during which run lengths would be set at 60 minutes but there

Week 10 Mar 1 to Mar 7   Codes: INTensive MINimal NORMal

Hour	Sunday	Monday	Tuesday	Wednesday	Thursday	Friday	Saturday
200	MIN	MIN	MS	MS	MS	NORM	NORM
	MIN	MIN	MS	MS	MS	NORM	NORM
400	MIN	MIN	MS	MS	MS	NORM	NORM
	MIN	MIN	MS	MS	MS	INT	NORM
600	MIN	MIN	MS	MS	MS	INT	NORM
	MIN	MIN	MS	MS	MS	INT	NORM
800	MIN	MIN	MS	MS	MS	INT	INT
	MIN	MIN	MS	MS	MS	INT	INT
1000	MIN	MIN	MS	MS	DOWN	INT	INT
	MIN	DOWN	MS	MS	NORM	INT	INT
1200	MIN	MS	MS	MS	NORM	NORM	NORM
	MIN	MS	MS	MS	NORM	NORM	NORM
1400	MIN	MS	MS	MS	NORM	NORM	NORM
	MIN	MS	MS	MS	NORM	NORM	NORM
1600	MIN	MS	MS	MS	NORM	NORM	NORM
	MIN	MS	MS	MS	NORM	NORM	NORM
1800	MIN	MS	MS		NORM	NORM	NORM
	MIN	MS	MS		NORM	NORM	NORM
2000	MIN	MS			NORM	NORM	NORM
	MIN	MS			NORM	NORM	NORM
2200	MIN	MS			NORM		NORM
	MIN						NORM

Week 11   Mar 8 to Mar 14   Codes: INTensive MINimal NORMAL

Hour	Sunday	Monday	Tuesday	Wednesday	Thursday	Friday	Saturday
200	MIN	MIN	MS	NORM	NORM	NORM	NORM
	MIN	MIN	MS	NORM	NORM	NORM	NORM
400	MIN	MIN	MS	NORM	NORM	NORM	NORM
	MIN	MIN	MS	NORM	NORM	NORM	NORM
600	MIN	MIN	MS	NORM	NORM	NORM	NORM
	MIN	MIN	MS	INT	INT	NORM	NORM
800	MIN	MIN	MS	INT	INT	NORM	NORM
	MIN	MIN	MS	INT	INT	NORM	NORM
1000	MIN	MIN	NORM	INT	INT	NORM	NORM
	MIN	MIN	NORM	INT	INT	NORM	NORM
1200	MIN	DOWN	NORM	INT	DOWN	NORM	NORM
	MIN	MS	NORM	NORM	NORM	NORM	NORM
1400	MIN	MS	NORM	NORM	NORM	NORM	NORM
	MIN	MS	NORM	NORM	NORM	NORM	NORM
1600	MIN	MS	NORM	NORM	NORM	NORM	NORM
	MIN	MS	NORM	NORM	NORM	NORM	NORM
1800	MIN	MS	NORM	NORM	NORM	NORM	NORM
	MIN	MS	NORM	NORM	NORM	NORM	NORM
2000	MIN	MS	NORM	NORM	NORM	NORM	NORM
	MIN	MS	NORM	NORM	NORM	NORM	NORM
2200	MIN	MS	NORM	NORM	NORM	NORM	NORM

A run schedule reflecting the inter-lab agreement. Two weeks' worth of running schedule is shown. MIN refers to a minimal tuning period. NORM refers to a normal tuning period. INT refers to an intensive tuning period. MS refers to machine studies time during which no experimental runs are scheduled. (P. Doing) would only be moderate constraints on beam position, and periods of "minimal tuning" during which the run lengths would be 90 minutes and the beam position would be held as steady as possible. This agreement was seen as the best way for the time being, for the particle group could properly pursue the goal of higher luminosity while the x-ray lab could be assured that during at least some periods the running of the machine would be run to their specifications. The head of CESR operations group and the head of the CHESS operations group were still working on a hardware feedback system, but no one knew quite what that system would look like. This agreement would see the lab through until that dreamt-of time when the technology would serve everyone's needs.

When I first learned of this agreement, I was disappointed. Being relatively new in the ranks of management, and not having lived through the early history of the laboratory, I felt as if we at the x-ray lab were giving something away that was rightfully ours. After all, weren't those periods already supposed to be for x-ray running? Why did we have experimenters scheduled for those times? Now we were saying that a certain amount of that time was going to be basically unusable and we were only getting in return usable time that was already ours. I remember how awkward it was when I first explained to users that some of their scheduled time would be unusable. I was in the minority, however. Others saw that we would now have to spend less of our time amassing technical arguments about how the machine performance and we could spend more time concentrating on the support and execution of x-ray experiments.

What was previously unsaid was now on the table for discussion. It was envisioned that the lab would now run more smoothly. A new line had been drawn. During intensive tuning periods, we at the x-ray lab could make no bones about what happened, and this allowed the particle group to have a legitimate chance to increase ring currents and therefore luminosity. Outside of those periods, however, the x-ray lab was to have a more powerful voice than ever.

"Hey, you guys haven't taken over yet?" I took the jestful barb in stride as I lay underneath a magnet assembly in the tunnel. "Well, its only a matter of time. Just remember to take care of us." My antagonist walked on, laughing. Just a passing joke, from one worker to another. "We'll see," I responded, going along with the line of the contact. I didn't really think about it, I just liked being involved in banter. He didn't know anything special. He was just talking. Nevertheless, things had definitely changed.

It had now been about four years since I first came out into this tunnel. We were no longer intruders. We belonged. I belonged. The x-ray lab was on the map. The crews knew it. They could sense the changes. Rumors were going around the lab. Everybody was talking about how the particle group had had to play the x-ray lab card hard in their last round of funding (Field Notes, Book 2, 3/28/97). Now they needed us! What was the future of this place? What was going to happen in ten years? Would the particle group still be funded? Did the future belong to the x-ray lab? I noticed that the machine guys seemed to be treating us x-ray lab guys a little differently. We had a job opening, and a lot of the machine technicians were interested in it. They were nicer to us. They started asking more questions about what we did. I felt more comfortable in the tunnel, more welcome. I liked walking around it, scanning the different shapes and textures of the various pieces of equipment. If I stopped and asked a machine person what a particular piece was, or how it worked, it wasn't seen as an interruption. I was growing more confident, I could talk the talk. They too had changed. They were more interested and more accommodating.

The Disappearance of Ion Trapping: Epistemic Politics and Laboratory Operations

If an anthropologist, or a new operator, or a user, or the director of the National Science Foundation walked into the lab after these negotiations were put in place and talked to people, read memos, looked at log books, and listened to announcements, she or he would not hear the term "ion trapping" in the intra-group discourse in the lab. In a stark change from when I started at the lab, the term "ion trapping" was no longer used, argued about, or deconstructed. It just wasn't there. For me, the disappearance of ion trapping followed several stages of understanding. Each stage of understanding goes along with a different model of the relationship between the technical and the political. I came to see technical statements as performances of models of the relationship between the technical and the political conducted with different audiences and forums of presentation in mind. As it was with the models of the relationship between codifiable theory and lived experience in the making of technical knowledge in the episodes between the operators and scientists described in the previous chapter, these models of technology are performed and put to use. But why do certain performances succeed and others fail, and how does this change over time? In the episode at hand, why did these different performances by the x-ray group and the particle group work at different times?

Ion Trapping as Technical Difficulty

When I began working as an x-ray lab operator, ion trapping was presented to me as an intractable technical problem that sometimes arose, and I took that to be so. It simply meant that the lab was experiencing technical difficulties. Referring to a model of unruly technology and helpless technologists, we choreographed our practice on the floor. I offered apologetic explanations to inexperienced users, who took what I said as a matter of fact. I told the experimenters that there was ion trapping in the machine and that we would refill immediately to correct the problem. At this point, the particle group operator and I were performing a certain model of the agency of technology. Sometimes, in some situations, technology simply breaks and there is nothing one can practically do about it. The only thing we could do was react to such times as best we could. As the machine operator spoke to me and I subsequently spoke to the x-ray lab experimenters, the image conveyed was that we were in one of those situations.

In order for the machine operator and me to perform the model of his passivity with respect to the inevitable breakdown of the machine successfully, we relied on a conception of our respective audiences as users who were not to question our presentations. To the machine operator, the x-ray group was simply a parasite that he was required by the rules of the lab to keep informed regarding the status of the machine. He regarded this necessity as a chore that kept him from his main business at the lab: running the machine. People at the x-ray lab were uninformed outsiders who were lucky to be able to use what they received. The accelerator was a "multiple-variable nightmare" that even physicists who had been working on it for more than a decade couldn't control. Why would the machine operator waste time trying to explain something that x-ray lab people lacked the capacity to understand? I (as a new operator) and the experimenters on the floor of the x-ray lab also saw ourselves as essentially passive. What could we do? Machine people ran the machine, x-ray lab people were informed of the status of the machine as a service provided by the machine group, and at this time the machine was experiencing technical difficulties. Referring to a model of unruly technology and helpless technologists, we choreographed our practice on the floor. I offered apologetic explanations to inexperienced users, who took what I said matter-of-factly. I told the experimenters that there was ion trapping in the machine and that we would refill immediately to correct the problem.

Ion Trapping as Operator Error

In time, an interpretation arose at the x-ray lab whereby the nature of a technical difficulty and the conception of the x-ray lab as a passive user of the machine changed. In time, there was seen to be a connection between the operators and the effect known as ion trapping. Like users who had been to the lab before, and like more senior staff members, I began to note an association between ion trapping and particular operator shifts. Ion trapping seemed to occur frequently on some shifts and infrequently on others. I began to take explanations of ion trapping differently. Ion trapping was not some inevitable technological occurrence; rather, it was an indication of a certain lack of talent on the part of the operator running the machine. Now we were dealing with operators who could control the machine if they were good, if they paid enough attention, if they were experienced enough, if they had enough skill. Tensions arose when the machine dumped and an explanation of ion trapping was offered. The x-ray lab experimenters and I were upset that the machine operator telling us that there was a problem did not have the ability to avoid the problem. The model of technical knowledge and identity had changed. The political and the technical were now connected to the extent that we felt that some control of the operation of the ring was possible. Ion trapping wasn't inevitable; the lab could do *something*. Machine operators should have more time on the machine before they were allowed to run the machine alone. Why couldn't there be an experienced operator on my shift!? In this milieu, users felt more of a right to protest and question what was going on. After all, if some operators could avoid ion trapping, why couldn't the others? The lab could arrange for better training of the CESR operators. Along with this model whereby problems were not inevitable came a new conception of ourselves as users. We could do something about our situation. After all, we knew best who did the job well and who didn't. Maybe the accelerator lab's management wasn't getting the right information. Knowing that there was a human element also meant that we could, and should, expect more from the machine operators. We became upset when ion trapping occurred. We asserted to the machine operator and among ourselves that such problems could be avoided. We did not see ourselves as passive users. Perhaps we didn't know how the machine worked, but we knew who *did* know and who *didn't*, and that was something. We had the right to expect the machine group to do what it took to avoid ion trapping.

While our conception of ourselves as more active users was growing, the machine operators' attitude toward us stayed the same as before. They informed us of the status of the machine without engaging us in explanations or discussions of the matter. As we came to see ourselves as properly involved in such discussions, the machine group's view of us as passive users remained the same, and tension rose on the laboratory floor. Conversations between x-ray operators and machine group operators were more heated, and x-ray lab experimenters voiced their frustrations more and more each time the machine had to be dumped because of ion trapping. This situation eventually became untenable, but not before I, and the x-ray group, underwent a profound shift in our conceptions of technical knowledge production and in our conceptions of who we were and who we could be as "users" of this laboratory.

Ion Trapping as Political Choice

After my initiation from the senior operator, as I moved up in the laboratory hierarchy, and as I spent more time in meetings away from the laboratory floor discussing episodes after the fact, I began to get a sense of a world in which a technical decision or explanation was not a simple matter. At the x-ray lab, we gradually became convinced that a policy toward the running of the machine was implicated in diagnoses like ion trapping. To us at this time, ion trapping was what happened when the machine was run in a certain way. To say that a run had to end because of ion trapping was to sidestep the fact that the machine had been *purposely* put into a configuration such that ion trapping would arise. Ion trapping was not a technical inevitability or a matter of lack of preparation, training, or skill. "Ion trapping" was code for a political choice. Here we performed a model of technical knowledge whereby the political and the technical were fundamentally intertwined. In the x-ray lab, we saw the machine group as adept choreographers of a complicated machine, able to draw on a complex repertoire of "technical" explanations to obscure their motivations and keep dissent at bay. This interpretation wasn't referred to explicitly outside our lab, either to the machine people or to the users, but it was at play in discussions among staff members of the x-ray lab, and it bolstered our subsequent actions. No one in our camp at that time protested that such a blurring of the technical and the political was possible (Field Notes, Book 7, 10/3/96). As we came to this conception, we also came to a new understanding of ourselves as users. We were no longer passive recipients of knowledge, nor were we a feedback mechanism whereby skill and talent were discerned. We were now, according to our conception of ourselves, active players in the reworking of the technical and political choreography at the lab. In this sense, we saw ourselves as aware simultaneously of the political and technical modes, working to perform a new order.

Ion Trapping and Epistemic Politics at the Lab

One author in the study of science and technology has written about how different conceptions of the agency of laboratory technology are put to use by laboratory members. Andrew Pickering (1995: 22–23) specifically argues that scientists sometimes perform a model of the nature of technology as passive and sometimes perform a model of technology as active in a dialectic of "resistance and accommodation" in a "mangle of practice" (ibid.: 104–109). My analysis draws upon Pickering's assertion but differs in that in Pickering there is a sense of inevitability, that scientists inevitably go through these stages as they work and that it is not something the practitioners can change and control. For this book, the question is who controls such dynamics and how. At the lab, we were putting these different models to use amid constraints of relations of access and voice with regard to authority and control. At the x-ray lab, we came to a conception whereby technology was permeated by politics and we were no longer passive recipients of knowledge or a feedback mechanism whereby skill and talent were discerned but rather active players in the reworking of the technical and political choreography at the lab.

With this model of technology as politics and a new sense of our-selves as active users added to our repertoire of justifications for action, we embarked on a project to change the way the machine was run. The difficulty was that the machine group did not explic-itly acknowledge our new understanding of the technology, and we could not do so explicitly either. We could not simply accuse them of ulterior motives; we knew that would be unproduc-tive. Our first strategy was to make it explicit that we were keep-ing records of the operation of the machine. By presenting our records in a matter-of-fact way to the machine group during the joint operations meetings of our groups, we worked to achieve our political goal of having the machine run differently. The machine group would engage with our discussion of particular problems during the "bad" periods, thus subtly and gracefully acknowledg-ing that "bad" running existed. In doing so, the machine group gave our group new and important epistemic space. We were recognized as able to monitor and record the operation of the machine using both outputs from our x-ray position monitors, which had recently undergone focused developments conducted jointly with the machine group, and readouts from experiments that were conducted on the x-ray lab floor. Although the machine group had its own beam position monitors, they allowed that the x-ray position monitors told the more precise story closer to the point of contact. The machine group, then, did not as a matter of course question the basis for our reports of machine opera-tions with respect to the stability of the beam in our areas. In the forum of the joint operations meeting, with the heads of both laboratories present amid high-ranking members of the techni-cal staff, it was granted that the technical basis for our reports was "our own business," so to speak. We were allowed to bring to the

table a technical parameter that would be accepted into considerations of the operation of the machine, and we were in charge
of the production of that technical parameter. Our status was not
simply passive recipients of "information" about the machine. In
this regard we were now accepted as knowledge producers. In this
way, through negotiation of a right to claim a different epistemic
territory, enacted particularly in the joint operations meeting,
a new dynamic of knowledge production and acceptance was
brought about at the lab whereby the x-ray group had a new voice
in the operation of the machine.

During the course of this change, the technical assessment of
ion trapping can be seen as taking on a new meaning at the lab.
For the machine group, ion trapping had originally been seen as,
and presented as, a risk of the normal operation of the machine.
At the x-ray lab, however, we saw ion trapping not as a risk of
normal operations but rather as the direct result of a purposeful
decision to conduct *in situ* experiments with the machine. It was
these *in situ* experiments that led directly to "technical" problems
like "ion trapping." We were working to have what was previously
considered the normal operation of the machine to be seen now
as abnormal. In time, a settlement was proposed. There would
be times when to run the machine would be "good" (for us) and
times when to run it would be "bad" (for us). We would agree on
these times in advance. Essentially, there would be times when
the machine was accountable to us and times when it was not.
At the x-ray lab, we saw this as doing away with the frustrating
practice of performing of the "technical" explanation of "ion trapping" as standing in for the political choice of how the machine
was run. The amount of "tuning" that was being conducted on
the machine was now the overriding factor in negotiations about

how the machine was performing. Announcements and discussions of ion trapping faded away. Now arguments between the two groups were over adherence to the run schedule, rather than over ion trapping and how best to "fix" it. As the x-ray group's conception of itself as a user had changed, and as the machine group came to accept the x-ray lab's more active and more controlling role, boundaries were renegotiated. What counted as a properly "technical" explanation by the machine group on the laboratory floor was resisted, and there was a new acceptance of "technical" presentations by the x-ray lab group at the joint operations meetings. Performances of old routines would no longer suffice as a new choreography of knowledge, identity, and authority was put in place.

Epilogue: Hardware Feedback as the Ultimate Solution

Throughout the episode described in this chapter, both groups promoted the vision of a hardware feedback system, a "technological fix" that would lay all consternation to rest. When it came time for me to leave the laboratory, a version of such a system was being "turned on" for the first time during running. I remember that hardware feedback was the primary topic of one of the last meetings of the x-ray operations group I attended. Everyone had been anxiously awaiting its implementation. The machine group's operations manager and the x-ray group's operations manager had been working together for more than a year, and a method of control had been agreed upon. The various operators had been briefed on the feedback system's operation, and over the weekend it had been turned on. The x-ray lab's operations manager was the last to arrive at the meeting, and everyone was anticipating his report. He explained that the system had some successes and some failures in its first run. For some of the time, the system ran as users experimented and the beam in the x-ray areas was very steady. Sometimes, however, the beams went the wrong way, causing the machine to dump. I remember the x-ray lab director listening silently to the operations manager's report and then exclaiming "Wait a minute, wait a minute. Did it go out of whack or did they *drive* it out of whack?"

5 The Absorption Correction: Biologists, Physicists, and a Re-Definition of Proper Experimentation

One day, at the lab, I was "tending to my garden"—an expression one of the scientists used to mean readying an experimental station for a run (Field Notes, Book 2, 2/19/97). I was weeding out some extra wires from the control rack and cleaning some papers off the desk. I didn't mind if a few cables were too long or too short, or if there were a few stray objects in the experimental hutch. Some guys had to have it perfect, but for me a little clutter was just right. I was the assistant operations manager. I spent a lot of time in meetings, planning and arguing, and I welcomed opportunities like this to work by myself out on the experimental floor. Just get it done, make it work; nothing else mattered but the here and now. There was an excitement to it—I would be doing experiments! I had been at the lab about five years now. A new kind of experiment involving protein crystallography, a new experimental way to map out the atomic structure of viruses and proteins, had become popular at my station. A staff scientist was in charge of the station, but he and I had been working together for a few years and I had been involved in the development of almost all the equipment used here. The experimental technique to be used tonight had already been published, but it was new enough that getting it to work was still a pretty big deal. For the scientist, who would not be an author on a paper published by the biologists, it was a "time sink"; he would work hard all night to get only, perhaps, an acknowledgement. The scientist pretty much left these nights to me, and I enjoyed them.

Protein crystallography experiments in general had always been prominent at the lab, but now it was really starting to take over. When the x-ray lab was first built, one station out of six stations was capable of doing these kinds of experiments, and even that one station supported other kinds of work too. When the lab expanded, two stations of nine were devoted to protein crystallography full time. Now, about half of the time on my particular station was given over to this kind of biological research, making it 2.5 out of 9. Among the physicists at the lab there was some grumbling about this trend. To them, the biologists were not "real" experimenters because what they really did was take measurements, not do conduct scientific experiments. The tone of this characterization would change, however, when the lab proudly announced in reports the various successes of the biologists in the journals *Science* and *Nature* magazines. To the us x-ray lab operators, all the groups who came to use the lab were users. That was the official word. And to us, users meant demands. Users meant pressure. Users meant having to explain things over and over. Users meant complaints. Users meant unreasonable expectations. Users just didn't understand. And the biologists were at the head of the user class in most or all of these categories. They understood the least about how a synchrotron operates, and they were the most unreasonable users. We were also aware, however, that there now existed large facilities (in the United States, the Advanced Photon Source at Fermilab in Chicago, the Stanford Synchrotron Light Source, and the National Synchrotron Light Source at Brookhaven National Laboratory on Long Island) that had many beamlines dedicated to protein crystallography from dedicated storage rings used solely as x-ray sources. In comparison with these sources, where runs could last more than 12 hours, ours was a short and sometimes bumpy ride. This tension between expectations built up at the other available sources and the realities on the x-ray lab's floor could lead to uncomfortable moments during the runs and was sometimes held against the biologists. On the night in question, I noticed that before the biology group was to arrive at my station there was a small block of time reserved by the new director of protein crystallography experiments at the lab, whom I had only recently met. I didn't know what he wanted to do. I was engrossed in setting up the station when he arrived with a colleague, whom I recognized as the head of a protein crystallography group at a university that was a regular user at the lab.

I was in the middle of checking out the functions of the monochromator and setting it to the proper wavelength when the director and his colleague arrived. I briefly explained to them that I was checking to make sure that the monochromator worked properly before setting it to the proper wavelength. As we talked, they asked me how short of wavelength the monochromator could be set to before the intensity of the x-rays provided to the station fell off significantly. I showed the x-ray lab's protein crystallography director a chart of wavelength versus intensity that was measured at the station a few years before, and he in turn showed it to the colleague. They wanted to know how it really was now, in practice. At how short a wavelength could this station be functionally run? When I told them it would be easy to check, they became noticeably more interested in my input to the conversation. They asked me how long it would take to do half a dozen intensity measurements at different wavelengths. When I told them it would

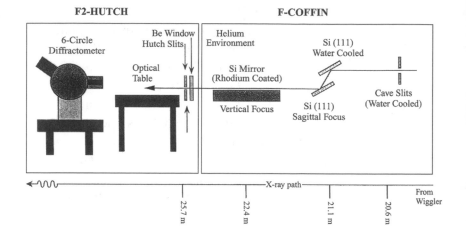

Schematic diagram of the F-2 x-ray station. "Hutch" refers to the room where the experiment is conducted. "Coffin" refers to a sealed-off vacuum chamber through which the x-rays pass on their way to the hutch. The monochromator is located inside of the sealed coffin. In this diagram, the monochromator is represented by the two thin rectangles marked Si(111). This refers to the fact that the monochromator uses two pieces of crystalline Silicon whose atomic plane orientation is (111) two monochromate the beam. Depending on the angle of the first monochromator crystal to the incoming "raw" synchrotron x-ray beam, only a particular wavelength of x-rays will diffract and be passed on. The second monochromator crystal diffracts this diffracted beam at a parallel angle such that the monochromatized x-rays proceed to the "hutch." (Doing 1994)

take less than an hour, they said "Let's do it!" For the next hour I could feel the excitement of the two researchers as I proudly put the monochromator through its paces and recorded the results (X-ray lab F2 Logbook). I was lost in my work, tending to my garden, and the hour passed in what seemed like five minutes. At the end, I remember feeling pleased as the director and his colleague thanked me and left. I was in a good mood as I turned my atten-

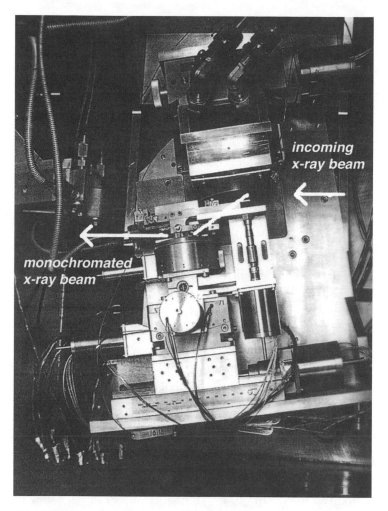

incoming
x-ray beam

monochromated
x-ray beam

The CHESS F2 x-ray monochromator in 2000. The whole mechanism can be rotated in order to "select" different wavelengths out of the incoming "white" x-ray beam and pass them on to the hutch. In this picture, the "first crystal" of the monochromator diffracts a monochromatic beam downward and at an angle onto the "second crystal," which then re-directs the beam (again through diffraction) to the hutch. A water cooling manifold can be seen entering the top of the "first crystal" assembly. The incoming beam is very powerful; the "first crystal" must be cooled so that it does not distort (due to a heat load), which would compromise the diffraction process. (P. Doing)

tion to the graduate students from the group scheduled for the night's run who had begun to mill about the station.

About two months later, as I was walking in the hallway, a staff scientist passed me and said "Hey, I heard you got author on a paper!" What was he talking about? My mind raced to think how this could be so. I went to check my mailbox. Sure enough, there was a reprint from a prominent journal in crystallography, and I was one of the five authors named (Helliwell 1993)! I quickly sat down and read the paper. The point of the paper was that perhaps protein crystallography experiments should be done at much higher energies, and that monochromators used for such work should be redesigned accordingly. My contribution, apparently, was to provide some hard numbers for the capability of our laboratory's monochromator systems. I remember being a little bit confused at the time. Other operators seemed to have differing views as to what kind of work was simply routine "set-up" and what kind of work was an important contribution to an experiment. In general, the other operators' view was that they did contribute in important ways and that those contributions were only labeled as routine by those in charge of the experiment. I remember thinking that in the case in question I hadn't done anything very different from the kind of work that had been taken for granted in the past. My thoughts were (1) that the new director of protein crystallography was trying to ingratiate himself with staff members and (2) that the other operators were going to get wind of this and then wonder if I was getting preferential treatment. At any rate, I was glad about what had happened.

A few weeks later, it was announced that the colleague who had helped conduct the short-wavelength experiment and write the paper was going to come to our laboratory to spend the summer in residence. With this news, considerations of a move to higher energies achieved the level of a slight buzz around the lab. It was a topic of conversations in offices, in hallways, and on the laboratory floor. I had learned by this time that the trick was to find out what was hot, then start working on it at just the right time. If you started on something too early, you were out of touch, wasting your time, and not doing what you were supposed to be doing. If you started too late, you were left behind and then all you ever got to do was what you were told to do. Even talking was tricky. If you spent too much time talking, you would be seen as not doing any work. On the other hand, if you didn't talk to people, you lost the pulse of the lab. If you managed it just right, you could get in on important early conversations, and with a well-placed test or measurement you could become a player in a new kind of development. Soon you could be an expert in a new area, licensed to indulge in conjecture about the future of the lab. I had already played a part in several monochromator developments, so it was acceptable for me to talk about monochromators in general with people. I kept my ears open for discussions of this topic, because I knew it was an area where I could be involved.

As the summer progressed, a controversy developed over whether the lab should consider converting some fixed-wavelength monochromators that were in use at the two stations dedicated to protein crystallography experiments to operate at shorter wavelengths, in line with the recent work at the third station. For the two fixed-wavelength stations, it wasn't simply a matter of adjusting the monochromator to a lower wavelength. Significant redesign of the experimental equipment and station would be required. Discussions about this took place in hallways, over lunch, and in offices. No official meeting was ever scheduled. No official presentation was ever made. But these informal discussions were very considered and went into quite a bit of detail. The arguments revolved around four main points and lined up along disciplinary boundaries. The biologists argued for the change and the physicists of the x-ray lab resisted. The following points were made.

Damage to Sample

Of the x-rays that impinge on a sample during an experiment, some are diffracted (into the pattern of interest to the experimenters), but some are absorbed by the sample. X-rays that are absorbed heat up the sample and degrade its structure. This effect is more prevalent at lower beam energies, where more x-rays are absorbed by a sample, than at higher beam energies, where more x-rays tend to pass through the sample. This problem was addressed by freezing the protein crystals in a cryogenic mist during data collection. For the biologists, this freezing was a big part of the technique. Ideally, data would be collected on one crystal, but crystals simply didn't last long enough in the beam. It was difficult to keep crystals frozen, and a mistake could easily render a crystal useless. Having crystals "die" in the beam was problematic even for well-frozen crystals, and often the lab's experimental staff played a large role in making this part of the experiment work. The lab even held special workshops to train outside users in this technique. The biologists saw in the move to high energies a chance to eliminate this whole part of the procedure, or at least render it far more forgiving.

The physicists typically did not see the freezing technique up close and in action. They were not involved in making this part of the experiment work. They knew freezing took a bit of work, but it was already routinely being done. Using more than one crystal for an experiment was not considered to be that much of a problem, since data analysis techniques were already in place to link data from one sample crystal to another. The physicists saw the biologists' freezing argument more in terms of what it said about biologists as experimenters than in terms of what it said about the biology experiment. To them, that the biologists didn't want to go through the trouble of freezing crystals was no reason to re-design entire experimental stations. That was just biologists wanting to eliminate anything that might hint at the difficult details of real experimenting.

Absorption Correction

When x-rays interacted with a sample, some of the x-rays were diffracted into a pattern of spots; others were absorbed by the sample. One had to take the effects of this absorption into account when considering the intensity of any diffracted spot. In other words, the intensity of any particular diffracted spot was modified by absorption as the x-rays traveled into and then out of the sample. To compensate for this effect, the standard practice was to take a calibration measurement of the sample. To the biologists, the absorption correction was a tedious and uninteresting aspect of data collection, requiring an extra measurement on each crystal and computer processing time in order to account for the differences between different crystals used for the same data set. At high energies, they argued, absorption was much less and absorption correction would be unnecessary. The physicists did not understand this line of reasoning. To them, practically every x-ray experiment had this "problem." Correcting for it was a normal part of x-ray work and a very simple matter. That the biologists would want to get around this fundamental experimental practice was, to the physicists, a reason to not take them seriously as experimenters. The physicists didn't feel that they should change a whole experimental set-up just so the biologists wouldn't have to do what is normally done in almost every other x-ray experiment.

Size of Detector

The detectors used to record the diffraction patterns were silicon-based charge-coupled devices (CCDs) Achieving the kind of sensitivity and dynamic range necessary for these kinds of experiments was difficult, and these detectors were expensive. They cost about half a million dollars each, and their size and price went up significantly as their sensitive area increased. For the crystallographers, this was problematic because more precise determinations of atomic structure required the recording of more diffraction spots—that is, larger diffraction patterns. At the time in question, the detectors could not record the full diffraction pattern required by most experimenters, and experimenters tended to live with the tradeoff of having less resolute answers or taking more time to piece together larger diffraction patterns from multiple "pictures." Proponents of the move to higher energies pointed out that using higher-energy x-rays in data collection shrunk the size of the diffraction pattern produced and therefore more data could be collected at any given time by the present detectors. The physicists noted that as the diffraction pattern grew smaller and the spots closer and closer together, at some point resolving individual spots becomes difficult. They also pointed to the fact that larger detectors were on the horizon. The problem could be fixed with larger detectors and the physicists didn't see this as a fundamental reason for changing the experimental setup.

Heat Load on Monochromator

One of the advantages claimed by the biologists for a move to higher energies involved an area of significant research and development in the lab referred to as the heat load problem. The heat load problem goes like this: the raw white light from the storage ring has power levels similar to those of an arc welder, meaning that almost any object hit by this beam will melt or otherwise be destroyed. Of course, in order to filter particular energies out of this beam, any monochromator used must intersect the beam. Further, in order to monochromate properly, the monochromator intersecting the beam must not have its atomic structure disturbed. Thus, a piece of crystalline material must intercept a beam containing the power of an arc welder and not only must not melt but must not even deform. This makes for a difficult engineering problem, and engineers and scientists has developed different and sometimes elaborate cooling schemes to solve it. Proponents of the move to higher energies pointed out that two interesting things related to heat load would occur if the energy were to be raised from 13 to 25 kilo-electron-volts (keV). First, the angle of the monochromator with respect to the beam could be shallower, thereby spreading out the beam footprint on the monochromating crystal and alleviating heat load. Second, and more important, since half of the power of the incoming beam was contained in the realm of energies below 23 keV and the other half contained in the realm of frequencies above 23 keV, and because the experiment would now be interested in only 25-keV x-rays, more than half of the incoming power could simply be blocked out by an absorbing device (such as an x-ray mirror or carbon filter) in front of the monochromator which would be set to only pass x-rays of 25 keV and above, thereby greatly reducing the heat load on the monochromator. The biologists argued that a move to higher energies was compelling because the "heat load problem," which was to them so intractable on these beamlines, could essentially be bypassed.

The physicists saw this issue differently. They saw the heat load issue as their own problem to work on and solve. Many x-ray physicists from a wide array of institutional settings had worked on different aspects of this problem for many years. Every year, workshops and even whole sessions at larger conferences were devoted to this topic. In the years before this episode, as biologists had come to rely heavily on the synchrotron data in their research, some tension had built up at the lab over this problem. As intense users of the x-ray lab's stations, when data were compromised due to heat load effects on the monochromators, the biologists had become more and more vocal in wondering why this problem hadn't been solved. I felt an element of pride and did not want to have this be a reason for altering an experimental technique. The physicists also pointed out that blocking out the heat load "upstream" limited future use of the station, as half of the incoming energies were now unavailable to potential experimenters

Forums for Performance

During the summer there were just a few public occasions where this issue was the topic of conversation, where arguments were laid out explicitly amid opportunities for questioning. The two forums were very different. The first was the x-ray lab's annual "User's Meeting"; the second was an international conference on synchrotron x-ray science. The Users' Meeting, held at the lab, was a forum for staff members from the lab to present the latest instrumentation developments at the lab and for scientists who had done research at the lab to present what were considered the most important results. This meeting lasted one day and generally had the aura of promoting the lab's accomplishments and direction for the future. The second forum was a several-day-long meeting of scientists and engineers from all over the world at which the latest scientific and instrumentation breakthroughs in the field of synchrotron x-ray science instrumentation were presented. The laboratory's users meeting occurred first. One section of this meeting was devoted to "user feedback." Typically that section was fairly short and was devoted to computer access, accommodations, and other support-related topics. This year, however, was different. Amidst the usual type of conversation, the visiting scientist on whose paper I was credited as an author raised his hand and was called on. "Why," he said, "don't we change the fixed energy crystallography monochromators to lower wavelengths? This lab is suited to the lower wavelengths, and there is a whole world of research to be opened up there, not to mention the fact that you bypass the heat load problem." The room fell silent for several seconds as the director of the x-ray lab tried to figure out what to do with this question in this setting. He was not entirely unprepared for such a discussion, as talk of this sort had been going on around the lab for several weeks, but he clearly felt that the question was inappropriate for this public setting. The question came off as antagonistic, which went against the general tone of the meeting. In response, the director said that there were many factors to consider and essentially postponed discussion of that topic until another time.

To me, the visiting scientist reinforced his outsider status by bringing up such a challenge in a setting where it usually was not discussed. This set an antagonistic tone and showed that he didn't understand that such decisions usually were "hashed out" informally with the lab staff before they were brought into more public settings. And, when they were brought into public settings, it was members of the synchrotron staff who did the bringing out, not "outsiders." Also, the tone of the question, a tone not lost on the lab director, implied that he had thought more about the technical details involved in this issue than the lab's director. Perhaps the visiting scientist felt that he wasn't getting anywhere with the informal approach and purposely took the opportunity to put the lab director on the spot. Whatever the motive or the level of awareness, tension was fairly palpable and a fruitful exchange did not occur.

A few weeks later, the issue received its most public airing at the international conference for the field of synchrotron x-ray experimentation held toward the end of the summer. At this meeting, the scientist who was visiting the x-ray lab for the summer was to give a presentation about the proposed move to higher energies. When the day came, there were at least 100 scientists and engineers in the audience, including the x-ray lab's director. For a personal reason, however, the visiting scientist who was to make the presentation was unable to attend the conference. A colleague (who admitted to the crowd at the outset that he wasn't that familiar with the details of the proposal) had to take over. The presenter fumbled through the points as I have outlined them. At the end, the x-ray lab's director was the first one with his hand in the air. In a loud voice he told the presenter that he just didn't hear any compelling reasons for redesigning the x-ray lab's protein crystallography stations experimental to operate at lower wavelengths. The fill-in presenter didn't say much in the way of response, and there was a bit of silence. A member of the audience then spoke up and said that perhaps the proposal should be put the other way around, that the x-ray lab's director should explain to the crowd why the experiments should stay at the present wavelength. To this the x-ray lab's director replied that a lot of history had gone into the experiment and it wasn't going to be moved for no reason. This pretty much ended the discussion and the presentation with a victory for the x-ray lab's director. The presenter, a soft-spoken man, didn't have a good grasp of the paper and couldn't really respond to the director's assertion. To a wide array of synchrotron researchers, it was clear that the director had not been seriously challenged. Had the scheduled presenter been there, I believe the tone would have been more confrontational and it would have been more apparent that a continuing argument was going on. The buzz in the room might have been different, and, as members went back to their respective labs to report, the issue might have gained some life. As it was, the presentation was a dud.

At the banquet for this conference, I shared a dinner table with the scheduled presenter who, while he had missed his presentation, had since arrived. Well, I thought, here was my opportunity! I could conduct an interview right here over dinner while the topic was hot and on everybody's mind. As the prime rib was being served, I told him in a somewhat offhand tone that I was pursuing a graduate program in the sociology of science as I worked at the x-ray lab, and that while I was working with him at the station, for example, I was also interested in the dynamics of laboratory work and in particular this controversy about the move to higher energies. I didn't know what to expect. How would he react? Perhaps he wouldn't want to talk to me. Maybe he would see my "ulterior motive" as reason for eliminating me from conversations about the project. Maybe he would just think it weird. I could see that he was paying attention as I spoke, a good sign. The sounds in the rest of the room became white noise as I explained my project to him.

To my surprise, he was not put off. In his country, sociology of science was a more prominent field than it was in the United States. He asked me if I knew a certain people, and I did; he then proceeded to profess his opinion on what was good about the field and what was bad. I listened politely. Then his tone changed as he began to explain to me what was going on in this particular controversy. Having spoken to the colleague who had given his presentation, he agreed that it had been quite disastrous. He also told me that he knew he had been "too blunt" at the Users' Meeting. He then explained to me that the problem was that the lab's director didn't understand how the network of experimentation and publication worked in his field. In his field, he explained, there was a time crunch that didn't exist in the x-ray laboratory director's own field. For the protein crystallographers working in this hot field, days mattered. He then started telling me about a molecule that a close colleague was working on. "What [the x-ray lab's director] doesn't understand," he told me, "is that if we can save time by not doing the absorption correction then we can scoop [a rival group]." He explained to me that in his fast-moving field, in which the stakes were becoming larger and larger, some groups weren't doing the absorption correction, even at longer wavelengths, and this has led to a controversy. "Now the purists aren't letting them publish," he told me, "so we can avoid the whole debate (and save time and get published)" by moving to shorter wavelengths. To the biologist, the physicist's stance was seen as coming from a field with less public interest and therefore less pressure. Physicists didn't understand that the biologists wanted to avoid the absorption measurement not because of a lack of ability, but rather because days mattered in the cycle of experimentation and publication in their field. They could do the absorption correction if they wanted to, but the time saved by working around it was crucial, and working at shorter wavelengths would allow them to bypass the absorption correction and get past journal reviewers without resistance on this point. To drive this point home, the biologist leaned over to me and said to me "OK, I'll tell [the x-ray lab's director] 'give us the beam time and we'll all measure the absorption corrections'

[and] he'll say 'there must be a better way to do this.'" The biologist told me that the main thing was that the x-ray lab's director was "is in a position of power" and therefore could write him off if he wanted. As I was listening intently and eating my chocolate mousse, a thump, thump, thumping rang out across the vast room as a distinguished-looking man in a neatly tailored blue suit and a pink tie tapped the head of a microphone. My table conversation was over. As all the heads turned toward the podium, I smiled at the visiting scientist thankfully. He nodded to me in response. After the speech, the tables broke up and I headed back to my hotel room to pour out the evening's conversation into my notebook (Field Notes, Book 4, 7/10/96).

After the conference, all went back to their respective labs, including the visiting scientist. At my lab, talk of the issue of redesigning the protein crystallography stations had died down considerably. When the issue was brought up in a staff meeting, it was quickly dismissed by the director, who succinctly stated that when he had found out that the main impetus was about avoiding an absorption correction he had written the whole thing off. Over the next year a number of protein crystallography experiments that mapped different aspects of different proteins were conducted at the tunable x-ray station at short wavelengths that bypassed the absorption correction, though at this time the fixed-wavelength stations remained fixed at their previously designed wavelengths.

6 Epistemic Politics and Scientific Change: The Rise of the Crystallographers

Protein crystallography is a field that claims to have the potential to revolutionize medical research. As the foundation of a technique referred to as "structure-based drug design" (meaning the designing of drugs based on measurements of the atomic structure of viruses and proteins), protein crystallography purports to give drug research a direct physical underpinning and thus to advance the field beyond trial-and-error techniques that do not analyze problems at the molecular level. In the late 1970s, when protein crystallography was first pursued at synchrotrons, the collection of x-ray data onto photographic film could take weeks and the digitization of the data from the film and the subsequent processing and analysis resulting in a map of the structure could take years. In the course of two decades, as more powerful synchrotron sources became available, new kinds of digital CCD (charge-coupled device) x-ray detectors were employed, and the time needed to solve a structure decreased steadily as supercomputers increased the power of data analysis exponentially. In the mid 1980s, collecting and processing data to solve a structure took years. By the mid 1990s, synchrotron data could be collected and a structure could be solved in a matter of months, and that time was getting shorter and shorter. The field considered

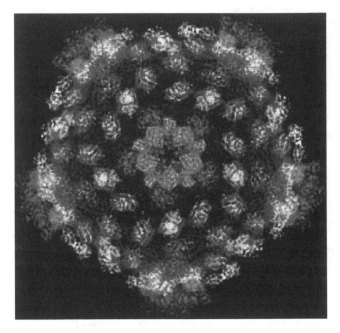

A processed visualization of a protein molecule. This image was built up from data collected at the x-ray lab. (Cornell 2000)

innovations that shortened the time needed to solve a structure significant, and there were often races to solve different structures.

As more and more structures were solved faster and faster, protein crystallography gained a significant measure of scientific and public acclaim and became a prominent aspect of research at the Cornell x-ray lab. The lab touted protein crystallography's potential when promoting itself to the public, and it cited the crystallographers' scientific successes in reports and proposals to the x-ray lab scientific board, to the university, and to funding agencies. In the 1990s, protein crystallography began to far outpace

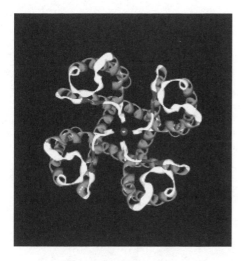

A three-dimensional ribbon diagram of an ion channel as described by Rod MacKinnon's group. This image was built up from data collected at the x-ray lab. (*Journal of Synchrotron Radiation* 11 (2004): 125–126)

the other kinds of work at the lab, and work done at the Cornell lab became prominent in the protein crystallography field as a whole. By 1999, more than half of the beam time allotted at the laboratory was for protein crystallography work and more than one-fifth of the "important" structures solved in the field were solved with data collected at the Cornell lab (Gruner 2003). In the years 1997–1999, crystallography work at the Cornell lab resulted in solved structures that were featured in 23 separate articles in *Science* and *Nature*. Very often, distinguished visitors to the university (including trustees) were brought into a special "theater" in which "bio-x" researchers could show off three-dimensional projections of molecules that had been solved. With its connections to the synchrotron and to the latest computing technologies, protein crystallography was readily placed at the heart of the biotech and information-technology revolutions. During this period, a new director for the x-ray lab was hired. Known for his work in developing the CCD detectors used in protein crystallography experiments, he had many connections to this rising field. The retiring director, whose work was in the general physics of x-ray matter interactions, had pushed the initial proposal for the Cornell x-ray laboratory in the 1970s and had played a major role in the build up of the lab from its beginning. After his arrival, the newly hired director wrote an article for a Cornell University newsletter in which he asserted: "The largest user community at CHESS [the Cornell High Energy Synchrotron Source] are the scientists who determine the atomic structure of proteins and viruses by x-ray crystallography. The biological revolution that is sweeping the world is based upon two technological developments: genetic engineering and protein crystallography. CHESS was centrally involved in the development of the cryo-freezing and x-ray detector methods that most protein crystallographers now use to determine molecular structures." (Gruner 1998: 3)

Cornell erected a new biotech building, and when President Hunter R. Rawlings III announced to the trustees that the university had received a grant of $25 million to support biotechnology research he received a standing ovation in the boardroom (Campi 2002). Around this time, time, the balance of funding at the x-ray lab also changed. In 1999, the lab's operating funds were supplemented by an "instrumentation grant" from the National Science Foundation for improvements to the experimental equipment associated with protein crystallography, including the monochromators used in the experiments. The amount of this instrumentation grant rivaled the entire operating budget of the x-ray lab for that year.

Meanwhile, the era of experimentation seemed to be coming to an end for the particle physicists at Cornell. In general, accelerator building in the United States had been stopped. The Superconducting Supercollider had been cancelled. Its smaller replacement, the Relativistic Heavy Ion Collider at Brookhaven, had been operational for a few years; its replacement, still smaller than the SSC, the Large Hadron Collider, was slated to come on line in 2008 but was based in Switzerland at CERN and was not a U.S.-led project. One of the last new facilities to be built in the United States, the B-Factory, had gone to Stanford instead of Cornell. In view of the Cornell machine's niche in B physics, not to be given the last B physics project was a death knell for experimental particle physics experiments at Cornell. With the loss of the B-factory, the latest round of funding for both the particle physics detector and the storage ring operation at Cornell was seen to have been contingent upon the rise in production and acclaim of the x-ray facility, which in turn was driven by the field of protein crystallography. At this time, it was generally seen as only a matter of time until the storage ring at Cornell would be dedicated solely to x-ray science.

Toward the end of my time working at the x-ray lab, I noticed many changes in how the biologists were portrayed and treated. I noticed, for example, that the lexicon of the lab had changed. There was much more of an acknowledged distinction between the x-ray facility and the protein crystallography group, and an assertion, bolstered by the now-known separate funding for experimental equipment, that the protein crystallography group was becoming a separate entity. The line between working for the x-ray group and the protein crystallography group was made more explicitly clear in interactions on the laboratory floor. The experimental stations where protein crystallography was done were commonly referred to as the "bio-x" stations. I also noticed that the experimenters and staff members associated with those stations were becoming more confident and more assertive. They expected detailed explanations regarding the equipment and the running of the machine, and they reacted with disdain if there was serious trouble. We operators were there to support them, and any accomplishments would be theirs. There were even jokes about the biologists taking over the lab.

This shift can be seen in announcements of the awarding of a prestigious prize in medical research for work done at the lab at this time. In 1999, Roderick MacKinnon, M.D., a professor at Rockefeller University and an investigator at the Howard Hughes Medical Institute, was awarded the Albert Lasker Basic Medical Award for his group's work on the role of ion channels in cellular interactions, some of which was done at the Cornell x-ray lab. It was known by laboratory members that about half of the recipients of the Lasker Award go on to receive the Nobel Prize, and MacKinnon's accomplishment was seen as the pinnacle of the important work that had been done in protein crystallography at the Cornell lab in the course of 20 years. The citation for the Lasker Award states that it was given to MacKinnon for "elucidating the functional and structural architecture of ion channel proteins, which govern the electrical potential of membranes throughout nature, thereby generating nerve impulses, and controlling muscle contraction, cardiac rhythm, and hormone secretion." In an editorial, the journal *Nature* said of MacKinnon: "He produced the first molecular description of an ion-selective channel. His 1998 paper in *Science* . . . on the crystal structure of potassium channels was a crucial step forward for this field and sparked much new research." (*Nature* 401, September 30, 1999: 414) *Science* cited the work on the structure of the ion channel as "one of the breakthroughs of 1998" and "the first physical characterization of the membrane protein responsible for the selective movement of [potassium ions] into and out of cells" (*Science* 282, December 18, 1998: 2158), editor in chief Floyd Bloom commenting: "After decades of wondering, electrophysiologists can now understand such riddles as how the potassium channel manages to keep out wrong ions, such as sodium, while shuttling an amazing 100 million potassium ions per second across the membrane. The structure reveals that the ions must pass through a narrow filter, where potassium ions fit snugly and briefly bind to the protein. The slightly smaller sodium ions cannot form this bond, making the filter an energetically unattractive place for them. . . . Membrane proteins are notoriously difficult to crystallize, but this year's triumph may prompt work on the thousands of other such proteins still waiting." At this time a last remnant of the old laboratory lexicon could be seen in the elevator that led to the main entrance of the lab: a plain typed note, taped up by the x-ray group's associate director, omitting the 'a' from MacKinnon's name and saying that he had won the award for work done at "x-ray" station F1. (Public announcements said "bio x-ray" station F1.) Four years later, the ultimate prize in science, the Nobel, was awarded to MacKinnon and two colleagues for this work.

I remember working with the students in Professor MacKinnon's group and with Professor MacKinnon during their different times at the lab. His group began to do work there during my last years there. It was interesting to me that some post-award accounts of the work give an "operator's-eye view" of MacKinnon's experimental work. I agree with the descriptions of him at the lab in these accounts. In 2004, the *Journal of Synchrotron Radiation* validated MacKinnon's presence and his practice at the lab as follows:

MacKinnon's SR (synchrotron radiation) experiments started from the use of CHESS in 1997. Needing powerful X-ray beams, when MacKinnon starting to knock on doors of synchrotron light sources, CHESS Director Sol Gruner offered him a rare allotment of Director's discretionary time to get the first X-ray measurements off the ground. Gruner, commenting on the happy occasion of MacKinnon's Nobel Prize said: "Little did Mac-Kinnon know that his first X-ray experiments in 1997 at the Cornell High Energy Synchrotron Source would lead to 30 more visits over the course of six years for a sum total of 1500 hours of X-ray beam time." The Mac-CHESS director, Dr Quan Hao, expressing his delight on recognition of MacKinnon's work said: "MacKinnon's style is well matched to the culture at CHESS where he participates in taking most of the X-ray data, working together with his students and postdocs. It is routine for a CHESS operator to find MacKinnon at the station collecting data through the night." (*Journal of Synchrotron Radiation* 11, 2004: 125) [Here "CHESS" or "Mac-CHESS" was used to describe the entire lab. —P.D.]

I too found him "well matched to the culture" at the lab. He was serious but not anxious. He was polite. He desired to be knowledgeable about the equipment at the lab without being too presumptively pushy. He was understanding of problems that we might have. He was easy to work with, and easy to work for.

Change in Practice

During the later part of my study, the biologists performed their new station at the lab in different ways. In the x-ray lab's weekly operations meeting, I now found myself on the other end of the kinds of assertions that we at the x-ray lab had been making in the machine group's operations meeting for the past few years. Each week, just after the opening report on the previous week's machine running by the x-ray lab's operations manager, the associate director of the bio-x lab would report on the successes and obstacles of the previous week's work on protein crystallography.

It used to be that I would report on the operations of the stations where protein crystallography work was conducted and he would report on data-collection aspects of the procedure, including crystal freezing, detector operation, and the experimental competence of the particular groups. Reports from the crystallography groups themselves were not trusted, and any statement from them with regard to station operation was not taken at face value but seen as requiring investigation. The saying was "You can't trust a crystallographer" (Field Notes, Book 2, 2/10/97). Now, however, the bio-x lab's associate director reported not only on the "experimental" aspects of the run but also on station operations, and reports from the crystallography groups were presented and were taken as carrying technical weight. I listened as problems with the station were reported. Though the reports were considered open to some questioning, the bio-x lab's associate director was considered technically competent to answer them. Just as the machine group now accepted the x-ray lab's reports of the position of the beam in the machine based on the x-ray lab's detectors, the x-ray lab now accepted reports of the x-ray lab's station operations by the bio-x group and the bio-x experimenters. I remember thinking that the tables had turned. We were not even really working together to solve problems; instead, the bio-x group was putting its expectations on the table, and it was up to us to meet them. At one particular meeting, the bio-x lab's associate director explained his reaction upon arriving at a station where his group was scheduled and finding that it was not operating properly. At the meeting, he slid a diagnostic scan that he had recorded onto the table and said: "When we saw that the crystal scan looked like this . . . why would we waste our time? We simply went home." (X-ray Lab F2 Logbook, 4/97) It was unheard of for a group to leave the lab during x-ray beam time that was scheduled for them and not return, and this was a strong assertion by the bio-x lab's director as to the expectations of the bio-x group and the consequences that would result from not meeting them. That the bio-x lab's associate director was not seen as "out of bounds" in this situation marks a shift in the dynamics of accountability at the lab.

Toward the very end of this period, I began working with a team of x-ray physicists at the lab to design a new monochromator for my experimental station, using money from the new bio-x instrumentation grant. As we were nearing the final stages of the design, and getting ready to order components from suppliers, the question of what wavelength range the monochromator should cover arose. One of the physicists asked me to email the associate bio-x director and ask him for the numbers. I did so. The associate bio-x director emailed me a one-sentence reply specifying the range, and we incorporated it in the design. I remember thinking how simple that was, and how uneventful. There was no argument by the physicists about who or what was really "right." The bio-x director was the authority on the matter. I remember thinking about how different things were now in comparison with the summer of controversy a few years before, when there was consternation, contestation, and argument. The routineness of this decision, once a site of agonizing, was held in place by the new epistemic-political relations at the lab between the biologists and the physicists, bulwarked by the rise of the crystallographers.

Rise of the Crystallographers

In the early stage of this study, the accepted division was that the physicists—"synchrotron jocks," as they referred to themselves— would provide monochromatic beam into the experimental station (meaning that the monochromator was their realm), and the protein crystallographers—biologists by training—would be in charge of preparing and maintaining their samples and would control the detection system used to record the diffraction pattern. At that time, the physicists saw the biologists essentially the same way they saw the operators, as users of information and equipment, and the same way the particle physicists saw the x-ray physicists. The biologists, like the operators and the x-ray physicists, did not share this view of themselves. The biologists came to characterize the physicists as an out-of-touch and old-fashioned group that did not understand the nature of their field with regard to experimentation and publication. Like the operators, and like the x-ray physicists in the machine group's meetings, the biologists appealed to a space to which the physicists had less direct access. Whereas the operators appealed to experience with the equipment, and the x-ray physicists appealed to their realm of experiments and their own position monitors, the biologists appealed to their own fast-paced world of experimentation and publication. In their network of experimenting and publishing, time was the most critical factor. If the absorption correction could be disregarded and they didn't have to mess with freezing and changing sample crystals during their experiments, then data processing could go faster, hurdles to publication could be overcome, and results, implying potential drugs, could be published sooner. Over the time period of my study, technical arguments were pushed by the biologists that were also claims for more control over the design and operation of a laboratory instruments that had traditionally been the domain of the x-ray physicists. The biologists made their challenges in the quasi-public settings of the lab's users' meeting and the international synchrotron instrumentation meeting in addition to the x-ray lab operations meeting and on the experimental floor. In all these different forums, amid shifts in the stream of funding to the lab and in the public awareness and scientific prestige that accrued to the protein crystallographers, their challenges became more accepted, less challenged as "un-technical" relative to the physicists' conceptions of proper experimentation. As time went on, they were granted epistemic space as spokespeople for "what had happened" with laboratory equipment and instrumentation, "what should be done" with regard to its operation and development, and how proper experimentation should be conducted.

With this epistemic re-ordering, after which the currents of accountability now flowed toward the biologists, a new emphasis on accounting and bureaucracy (implicated in all the episodes described in this book) arose as the laboratory developed away from the informal looseness of the early years. The new kinds of reporting and justifications for actions now prevalent at the lab were such that the operation of the machine was altered via the proxy assertions of the x-ray physicists to better accommodate the biologists experimental practice, reports about the operation of experimental equipment were judged by the touchstone of the biologists' experimental techniques, and experimental methods were validated on the basis of time limits and the publishing dynamics of the protein crystallography field. This realignment of the laboratory toward biology was realized through specific negotiations of epistemic politics in the various forums available to laboratory members. Indeed, "the laboratory" can be thought of as a negotiated collection of forums with different linkages of access and voice that make possible new expressions of authority and control in scientific practice such that traditional categories of "inside the laboratory" and "outside the laboratory" break down. Change occurred through new negotiations of epistemic politics throughout these recursively related forums.

Over the seven years of my study, the general realignment toward biology in science, and toward protein crystallography in synchrotron science, found expression at this laboratory as the dynamics of epistemic politics there changed and endured such that important scientific resources came to be controlled and defined by the protein crystallographers, thus, in turn, bolstering the ascendance of their field. Summarizing this overall shift of the lab toward biology, the x-ray lab's original director, upon returning from a meeting with National Science Foundation representatives in Washington, assessed the future of the lab in a matter-of-fact way: "Physics [had] prospered in the service of, shall we say, workfare with the military thing. But the military budget pales in comparison to medicine." (Field Notes, Book 2, 1/27/97) The biologists at the lab were accounted to in ways that changed and endured through new assertions of epistemic politics in the recursively related forums of authority and control that constituted the Cornell Synchrotron laboratory over the years 1993–2000.

Contingency Revisited

How is this enduring change in scientific practice implicated in the status of the facts that emerged "from," and were used "in," the lab? In this book, I have worked to build upon the pioneering constructivist laboratory studies to give a compelling account of how the criteria for judging scientific facts might change and endure in practice while following Mike Lynch's admonition to not be preoccupied with relying on 'antecedent variables' that impinge on the action of practice from the outside. It is in this sense that I invite the connotation from the title of the book of Czechoslovakia's well-known Velvet Revolution. I am not asserting any kind of direct analogy to that historical event; rather, I am invoking the general idea that, in my account, change was not "top down" and work was required on the part of the people and groups involved to assert an maintain newly accepted ways of knowing that belied the seeming effortlessness of the regime change. In this account, part of this change involved the very criteria involved in establishing scientific facts. This is an advance on the lab studies analyses that have come before in that this account provides a fleshed-out means of understanding how such a change can be made to gain traction and endure in practice. But the question still remains: Does the fact that such change occurred mean that scientific facts have been shown to be contingent on practice? In considering the actions described in this book, this question must be face head on yet again.

We have seen that status of the absorption correction as a necessary criterion of a fact claim with regard to the structure of a protein underwent a change. A new model of proper experimentation in which bypassing the absorption correction in the experimental collection of x-ray crystallography data was *not* seen as a reason to discount an experimental claim about the structure of a macromolecule was put in place and made to endure through the rise of the field of protein crystallography. If the changing treatment of the absorption correction in experimental and publishing practice were to be implicated in the epistemic status of fact claims arising from that practice, then it could be said that contingency had been traced from the ground up, from contingent practice into *enduring* scientific fact. Here is where my conclusion departs from the pioneering lab studies. Even though the absorption measurements (previously required) were shifted out of the criteria for fact judging with regard to claims about the structure of the proteins through an epistemic-political regime change, that in itself does not define the facts that arose from that new regime of practice as contingent effects. The absorption of x-rays into the kinds of materials and crystals used for protein crystallography experiments *could* be measured, if the time and resources were allotted to doing so. The accepted judgment that it need not be done because the relation between x-ray energy and absorption into the types of materials used was understood enough to consider it negligible does *not* define those facts that did not rely on the measurement to be contingent effects of practice. While such facts may indeed be impugnable, they would be so on other grounds than the fact of such a judgment being used—namely, that the judgment might turn out to be wrong.

The same can seen to be true for the other knowledge claims described in this book. In the first chapter, the scientist who relied on my "bodily sensing ability" to gauge whether a silicon crystal was clamped properly took that crystal, after adjustment, to the x-ray beamline, where the proof was in the pudding in that when the x-ray beam diffracting off of the crystal looked and registered in intensity, angle spread, and wavelength spread the same as previous beams off of previous "undisturbed" crystals, the adjustment could be compellingly said to have worked. With the vacuum pumps and the question of whether any operators had "done the laying on of the hands thing," again, measurements of current, voltage, and pressure in the chamber, among other measurements, could be used to distinguish the particular pump under question from other working pumps, or not. With the ion trapping in the second chapter, whether and how the synchrotron was put into particular configurations such that ion trapping would occur was up for debate, but that it occurred— that ions used by the pumps to draw out particles from the storage ring and thereby create a vacuum were interacting with the electrons and positrons in the ring in such a way as to disrupt the trajectory of the electrons and positrons in the ring—could be cross-checked through various position monitors, temperature gauges, x-ray output monitors, and pressure gauges in the same sense. My analysis in this book *and* those of the pioneering lab studies do not impugn scientific facts as contingent effects in principle.

Here is the crucial way that laboratory studies can contribute to considerations of technical expertise. We have seen that control of epistemic territory can be negotiated in the course of asserting technical knowledge claims. Indeed, we have seen that a registration between different modes of authority and control and different understandings of the epistemic basis for technical claims was necessary for claims to be accepted as justifications for action. We have seen that the "outside" and the "inside" of the lab are not stable, given categories; practice pervades them, as any reconstruction of where scientific "work" is done is a *post hoc* production.

The laboratory members offer us lessons, then. As we follow them, we can see how epistemic politics can re-define the criteria for reality claims, and how that re-definition can endure. Anyone seeking to rely on or to analyze technical knowledge must be attuned to such dynamics. But this is not to say that potential coercions of reality cannot be checked. Practically, it's a matter of resources, voice, and access to do such checking. Indeed, an understanding of how epistemic politics can wrongly coerce reality, such that resources should be put to the task of "un-coercing" it, is part and parcel of the job of a science studies ana-lyst! This book has described episodes where such resources and access were renegotiated and put to use in new, enduring ways with regard to technical and scientific assertions amid chang-ing relations of authority and control at a present-day hybrid biology-physics laboratory. The terms of reportage of the techni-cal did change, and these changes were implicated in the rise of protein crystallography as the most important field in synchro-tron science amid the rise of biology in science in general. But while the assertions of knowledge-making ability put to use in practice at the lab were expedient coercions of accepted reality that endured, this coercion itself was not implicated, in the end, in the ultimate status of fact claims that were produced "in" and emanated "from" the lab. There *were* fundamental changes in the epistemic-political fabric of scientific practice, but that change was *not* implicated in the epistemic status of the products of that practice. The revolution was velvet.

The lab had changed. My life was changing. I could feel myself disen-gaging from the fray. In the months leading up to my departure from the lab, I wondered who I really was and what I was really doing. Only at the end, when I was most familiar with the scene, did I feel like an interloper. I remember the day I left the lab. The high-pressure equip-ment that was waiting to replace me and my desk in the prime real estate of my office right off of the experimental floor loitered outside my door as if embarrassed in front of the voluptuous surveillance of the Georgia O'Keefe print gazing out over the hallway. Everything was strange again. People with faces I didn't recognize walked the halls. Unfamiliar characters were focused on jobs, sounds, and problems that I didn't understand. I felt like a spirit among them. I shook hands. I smiled. I tried to explain about graduate school, philosophy, and history. I received accommodating smiles in return. Old arguments dissipated. Tensions faded. I walked around the experimental floor as if to say goodbye to my old equipment. I looked up at the ceiling, and my eyes drifted over to the enormous gouge in the wall marked by the sign reading "Top quark went through here." I looked at the giant girders jutting out of the ceiling crane aimed right at the gouge and smiled—it was a funny joke.

As I stepped into the elevator to leave, I could see the blinking lights of the control room and hear the gentle chiming of the synchrotron bell as the doors glided shut and, with a gentle jerk, the car began to rise.

References

Anderson, Warwick. 1992. The reasoning of the strongest: The polemics of skill and science in medical diagnosis. *Social Studies of Science* 22: 653–684.

Barley, Stephen, and Beth Bechky. 1993. *In the Back Rooms of Science: The Work of Technicians in Science Labs*. National Center for Educational Quality in the Workforce.

Barnes, Barry. 1882. The science-technology relationship: A model and a query. *Social Studies of Science* 12: 166–172.

Batterman, Boris. 1986. Making good use of synchrotron radiation: The role of CHESS at Cornell and as a national facility. *Cornell Engineering Quarterly* 20 (4): 2–13.

Batterman, Boris, et al. 1977. Final Report: Workshop on the Application of Synchrotron Radiation to X-ray Diffraction Problems in Materials Science. Cornell University, July 27–29.

Brown, Laurie, Max Dresden, and Lillian Hoddeson, eds. 1989. *Pions to Quarks: Particle Physics in the 1950s*. Cambridge University Press.

Callon, Michel. 1987. Society in the making: The study of technology as a tool for sociological analysis. In *The Social Construction of Technological Systems: New Directions in the Sociology and History of Technology*, ed. W. Bijker et al. MIT Press.

Campi, Esther. 2002. Cornell bio center gets $25M. *Ithaca Journal*, November 2.

Cardwell, Donald. 1976. Science and technology: The work of James Prescott Joule. *Technology and Culture* 17: 674–687.

Cathcart, Brian. 2004. *The Fly in the Cathedral*. Viking.

CESR. 1977. CESR, An Electron-Positron Colliding Beam Facility: A Proposal to the National Science Foundation. Cornell University. Submitted to NSF in October.

CESR DR. 1977. Design Report, An Electron-Positron Colliding Beam Facility. Cornell University. Submitted to NSF in December.

CHESS. 1977. Proposal to Establish a High Energy X-Ray Synchrotron Radiation Laboratory Associated with the CESR 8-GeV Storage Ring. Submitted to NSF September 30.

Collins, H. M., and Steven Yearley. 1992. Epistemological chicken. In *Science as Practice and Culture*, ed. A. Pickering. University of Chicago Press.

Collins, H. M. 1985. *Changing Order: Replication and Induction in Scientific Practice*. Sage.

Cornell University. 2000. *The Science at Wilson Laboratory*.

Cowan, Ruth Schwartz. 1996. Technology is to science as female is to male: Musings on the history and character of our discipline. *Technology and Culture* 37 (2).

Doing, Park, Jeff White, and Qun Shen. 1994. A high-power and high-flux x-ray wiggler station at CHESS. *Nuclear Instruments and Methods in Physics Research A* 347: 73–76.

Doyle, D. A., J. M. Cabral, R. A. Pfuetzner, A. L. Kuo, J. M. Gulbis, S. L. Cohen, B. T. Chait, and R. MacKinnon. 1998. The structure of the potassium channel: Molecular basis of K^+ conduction and selectivity. *Science* 280: 69–77.

Faulkner, Wendy. 1994. conceptualizing knowledge used in innovation: A second look at the science-technology distinction in industrial innovation. *Science, Technology, and Human Values* 19 (4): 425–458.

Galison, Peter. 1997. *Image and Logic: A Material Culture of Microphysics.* University of Chicago Press.

Gamber, Wendy. 1995. "Reduced to science": Technology and power in the American dressmaking trade, 1860–1910. *Technology and Culture* 36 (3): 455–482.

Gruner, Sol. 2003. CHESS: A bright and penetrating light. *Arts and Sciences Newsletter* (Cornell University) 24 (1): 4–5.

Hartman, Paul. 1988. Some early synchrotron radiation history. Presented at CHESS Users Meeting, 1988.

Heilbron, J.L., and Seidel, Robert. 1989. *Lawrence and His Laboratory: A History of the Lawrence Berkeley Laboratory, Volume 1.* University of California Press.

Helliwell, J., S. Ealick, P. Doing, T. Irving, and M. Szebeny. 1993. Towards the Measurement of Ideal Data for Macromolecular Crystallography Using Synchrotron Sources. *Acta Crystallographica D* 49: 120–128.

Hoddeson, Lillian, et al., eds. 1997. *The Rise of the Standard Model.* Cambridge University Press.

Keller, Evelyn Fox. 1983. *A Feeling for the Organism: The Life and Work of Barbara McClintock.* Freeman.

Kevles, Daniel. 1995. Preface: The death of the Superconducting Super Collider in the life of American physics. In *The Physicists*, ed. D. Kevles, third edition. Harvard University Press.

Kevles, Daniel. 1997. Big science and big politics in the United States: Reflections on the death of the SSC and the life of the Human Genome Project. *Historical Studies in the Physical and Biological Sciences* 27: 269–298.

Kline, Ron. 1995. Construing 'technology' as 'applied science': Public rhetoric of scientists and engineers in the United States, 1880–1945. *ISIS* 86: 94–221.

Knorr Cetina, Karin. 1981. *The Manufacture of Knowledge: An Essay on the Constructivist and Contextual Nature of Science.* Pergamon.

Knorr Cetina, Karin. 1995. Laboratory studies: The cultural approach to the study of science. In *Handbook of Science and Technology Studies*, ed. S. Jasanoff et al. Sage.

Knorr-Cetina. Karin. 1999. *Epistemic Cultures: How the Sciences Make Knowledge*. Harvard University Press.

Konig, Wolfgang. 1996. Science-based industry or industry-based science? Electrical engineering in Germany before World War I. *Technology and Culture* 37 (1): 71-92.

Krane, Kenneth. 1983. *Modern Physics*. Wiley.

Kuhn, Thomas. 1962. *The Structure of Scientific Revolutions*. University of Chicago Press.

Latour, Bruno, and Steve Woolgar. 1979. *Laboratory Life: The Social Construction of Scientific Facts*. Sage.

Law, John. 1987. Technology and heterogeneous engineering: The case of Portugese expansion. In *The Social Construction of Technological Systems: New Directions in the Sociology and History of Technology*, ed. W. Bijker et al. MIT Press.

Layton, Edwin T. 1976. American ideologies of science and engineering. *Technology and Culture* 17: 688-701.

Lelas, Srdjan. 1993. Science as technology. *British Journal for the Philosophy of Science* 44 (3): 423-442.

Lynch, Michael. 1985. *Art and Artifact in Laboratory Science: A Study of Shop Work and Shop Talk in a Research Laboratory*. Routledge & Kegan Paul.

Mayr, Otto. 1976. The science-technology relationship as an historiographic problem. *Technology and Culture* 17: 663-672.

Merton, Robert. 1957. *Social Theory and Social Structure*. Free Press.

Molella, Arthur, and Nathan Reingold. 1991. Theorists and ingenious mechanics: Joseph Henry defines science. In Reingold, *Science, American Style*. Rutgers University Press.

Oudshoorn, Nelly, and Trevor Pinch, eds. 2003. *How Users Matter: The Co-Construction of Users and Technologies*. MIT Press.

Rice, Dave, and Ernest Fontes. 1999. X-ray vision is real! On tour at CHESS. *Connecting with Cornell: News From the Office of the Vice Provost of Research* 13 (3): –.

Patterson, Richie. 2002. Elementary particle physics. *Cornell University Arts and Sciences Newsletter* 24 (1): 1.

Pickering, Andrew. 1995. *The Mangle of Practice: Time, Agency, and Science*. University of Chicago Press.

Pickering, Andrew. 1984. *Constructing Quarks: A Sociological History of Particle Physics*. University of Chicago Press.

Pinch, Trevor. 1997. Kuhn: the conservative and radical interpretations. *Social Studies of Science* 27 (3): 465–482.

Pinch, Trevor. 1986. *Confronting Nature: The Sociology of Solar Neutrino Detection*. Reidel.

Popper, Karl. 1963. *Conjectures and Refutations: The Growth of Scientific Knowledge*. Routledge and Kegan Paul.

Saulnier, Beth. 1996. Real time beam time. *Cornell Engineering Magazine* 2 (1): 6–14.

Schweber, S.S. 1992. Big Science in Context. In *Big Science: The Growth of Large Scale Research*, ed. P. Galison and B. Hevly. Stanford University Press.

Shapin, Steven. 1989. The invisible technician. *American Scientist* 77 (6): 554–563.

Staudenmaier, John. 1985. *Technology's Storytellers: Reweaving the Human Fabric*. MIT Press.

Vincenti, Walter. 1990. *What Engineers Know and How They Know It: Analytical Studies from Aeronautical History*. Johns Hopkins University Press.

Whalley, Peter, and Steve Barley. 1997. Technical work in the division of labor: Stalking the wily anomaly. In *Between Craft and Science: Technical Work In U.S. Settings*, ed. S. Barley and J. Orr. ILR Press.

Index

Inside Technology

edited by Wiebe E. Bijker, W. Bernard Carlson, and Trevor Pinch

Paul N. Edwards, *The Closed World: Computers and the Politics of Discourse in Cold War America*

Herbert Gottweis, *Governing Molecules: The Discursive Politics of Genetic Engineering in Europe and the United States*

Joshua M. Greenberg, *From Betamax to Blockbuster: Video Stores and the Invention of Movies on Video*

Kristen Haring, *Ham Radio's Technical Culture*

Gabrielle Hecht, *The Radiance of France: Nuclear Power and National Identity after World War II*

Kathryn Henderson, *On Line and On Paper: Visual Representations, Visual Culture, and Computer Graphics in Design Engineering*

Christopher R. Henke, *Cultivating Science, Harvesting Power: Science and Industrial Agriculture in California*

Christine Hine, *Systematics as Cyberscience: Computers, Change, and Continuity in Science*

Anique Hommels, *Unbuilding Cities: Obduracy in Urban Sociotechnical Change*

Deborah G. Johnson and Jameson W. Wetmore, editors, *Technology and Society: Building our Sociotechnical Future*

David Kaiser, editor, *Pedagogy and the Practice of Science: Historical and Contemporary Perspectives*

Peter Keating and Alberto Cambrosio, *Biomedical Platforms: Reproducing the Normal and the Pathological in Late-Twentieth-Century Medicine*

Eda Kranakis, *Constructing a Bridge: An Exploration of Engineering Culture, Design, and Research in Nineteenth-Century France and America*

Christophe Lécuyer, *Making Silicon Valley: Innovation and the Growth of High Tech, 1930–1970*

Pamela E. Mack, *Viewing the Earth: The Social Construction of the Landsat Satellite System*